ALDOUS HUXLEY'S HANDS

ALDOUS HUXLEY'S HANDS

HIS QUEST FOR PERCEPTION AND THE ORIGIN AND RETURN OF PSYCHEDELIC SCIENCE

ALLENE SYMONS

Prometheus Books

59 John Glenn Drive
Amherst, New York 14228

Published 2015 by Prometheus Books

Aldous Huxley's Hands: His Quest for Perception and the Origin and Return of Psychedelic Science. Copyright © 2015 by Allene Symons. All rights reserved. No part of this publication may be reproduced, stored in a retrieval system, or transmitted in any form or by any means, digital, electronic, mechanical, photocopying, recording, or otherwise, or conveyed via the Internet or a website without prior written permission of the publisher, except in the case of brief quotations embodied in critical articles and reviews.

Cover design by Jacqueline Nasso Cooke and Stephanie Starr
Image of Aldous Huxley's hand by Howard Thrasher (the author's father)

Trademarked names appear throughout this book. Prometheus Books recognizes all registered trademarks, trademarks, and service marks mentioned in this book.

Inquiries should be addressed to
Prometheus Books
59 John Glenn Drive
Amherst, New York 14228
VOICE: 716–691–0133
FAX: 716–691–0137
WWW.PROMETHEUSBOOKS.COM

19 18 17 16 15   5 4 3 2 1

Library of Congress Cataloging-in-Publication Data Pending

ISBN 978-1-63388-116-7 (paperback)
ISBN 978-1-63388-117-4 (ebook)

Printed in the United States of America

CONTENTS

PART I—THE WORLD'S BIGGEST DRUGSTORE

PART II—ONE BRIGHT MAY MORNING

PART III—AFTER ALDOUS

PART I

THE WORLD'S BIGGEST DRUGSTORE

CHAPTER 1

ADAPTATION

I am glad curiosity brought you here, and I imagine it's because *Brave New World* has become a kind of shorthand, a catchphrase for technology overtaking humanity.

Aldous, whose own curiosity took him in wildly experimental directions, lives on in the scrapbook of a generation. He looks out at us from the album cover of the Beatles' *Sgt. Pepper's Lonely Hearts Club Band*, a cameo-nod to how Huxley laid groundwork for the psychedelic revolution with his slim book *The Doors of Perception*, which in turn inspired a 1960s rock band called The Doors. A dozen years later, the same year as the band's Jim Morrison died at age twenty-seven, Huxley's story of a documented case of possession and exorcism, called *The Devils of Loudun*, became a mind-searing Ken Russell film. Such impressions linger.

Huxley died fifty years ago almost to the day as I write this, but he still holds many of us in a kind of retro-cultural thrall. My own take, however, is personal.

During the 1950s when I was a child, Aldous and Maria Huxley hosted weekly salons at the Huxleys' home in Hollywood on North Kings Road, gatherings my father often attended, and sometimes my mother came along too. Their Tuesday night group participated in séances and age-regression hypnosis, sought ways to prove ESP, and explored techniques to bring about remote healing. Such exploration of the byways of the mind led to experimentation with substances that later came to be called psychedelic.

My father was invited to join this eclectic circle of friends after Aldous heard about my dad's amateur photographic study of human hands, which at the time amounted to hundreds of hands of musicians, twins, stutterers, proclaimed psychics, and schizophrenic patients.

At the time I write these introductory pages, psychedelic substances forbidden for decades are back in clinical trials. Fingerprints and palm prints are making headlines because of studies underway in fields ranging from medicine to global security, all in pursuit of an incorruptible wax seal to verify identity. Soon we may be known less by our passwords and strings of other numbers than by subtle patterns found in our skin.

You might say old is new, or new is old, or that the leading edge oddly resembles shamanic and occult arts with branching histories reaching back millennia.

Which brings me to the heart, or rather the hand, of this story and the common bonds between the Tuesday circle, my father, and Aldous.

I'll start at the end then go back to the beginning. As final exits are concerned, Huxley's death stands out as a unique act, also for the historic day on which it occurred. Around noon on November 22, 1963, Aldous scribbled with some difficulty a note requesting an intramuscular injection of lysergic acid diethylamide, or LSD-25. Around the same time in Dallas, Texas, an open black limousine carrying President John F. Kennedy, his wife, Jackie, along with Texas Governor John B. Connally and his wife, continued along the prescribed parade route and neared a curve.

In distant Los Angeles, inside a Spanish-style stucco house under the shadow of the Hollywood sign, a frail Aldous Huxley neared the end of his three-year struggle with laryngeal cancer. The LSD presumably took effect while the nursing staff and the doctor, along with a few friends and family members, attended the author or waited in a room nearby. On an ordinary day he would have been writing in his bright upstairs study, where a reproduction from the Edgar Degas bather series hung on the wall. In that room he had hoped to continue his life's work: the poems and essays, short stories and articles, biographies, plays and screenplays, and a dozen novels, most recently *Island*, more famously *Brave New World*.

During the last weeks of his illness, while the satisfaction of making decisions remained within reach, Huxley managed to complete at least three tasks. One was finishing an article called "Shakespeare and Religion." Another, it appears, was arranging to have a photograph taken of his right hand. Back when Aldous was a decade younger, my father had captured a similar image of both of his hands, but this new black-and-white photo seems to have been taken in October or early November.

When the photographer arrived at the Huxley home he must have wondered, *Why does this famous writer want a picture of his hand, and why now?* The photographer used standard equipment, unlike a decade earlier when my dad brought his custom-designed camera rig with its specially angled strobe lights. This second photographer did not control for glare. His flash washed out details of an intriguing area: a triangle-like configuration, a rare but significant indicator sometimes found where one of the fingers joins the palm.

Huxley's last task took place on November 22, after Laura Huxley and the attending physician saw a change in vital signs. Scrawling the last line he would ever write, Aldous asked for an intramuscular dose of lysergic acid diethylamide. In May of 1953 he had experimented with a similar substance called mescalin (or mescaline) and now, a decade later, the physician gave his tacit consent.[1] Huxley's last wish was granted.

Who knows what a person perceives in the transition called dying, whether JFK felled by a rifle in Dallas or Aldous Huxley under the spell of LSD. This time Huxley would write no guidebook like *The Doors of Perception*, no map showing readers what lies ahead in the valley of mystics and madness.

The decade-long task of writing this book, far longer than I had imagined, began one day when I was poking around in the garage behind my grandparents' 1910 Craftsman bungalow. Long Beach had been their final destination after migrating from Iowa to Los Angeles County back in the 1920s, and ever since then this garage between my grandparents' larger house and our smaller one had sheltered three generations of my family's cars and castoffs.

I was wearing rubber gloves and tossing out stuff to make more room for cartons of books I'd previously hauled home after a decade of living in Manhattan. The day before this I'd reached deep into a cubby hole and pulled out an intact Art Deco Roseville Iris vase, a valuable find, but on this day no such booty, at least not so far. By now the midday heat had released a stifling potpourri of dust and old redwood siding.

Then something caught my eye in the gloom, and I saw a jumble of cardboard boxes peeking out from behind a row of rusty paint cans. Was this trash or treasure? The first clue was their labels, written in my father's distinctive draftsman's style. Four were labeled "The Hand." One was labeled "The Church."

It dawned on me that the first group contained my father's unpublished study and its supporting photographs, the yield of his seven-year project of the early 1950s that, like in the Harrison Ford film *Mosquito Coast*, in many ways held our family hostage to his obsession. At the time, I didn't feel much like a hostage because when I was a kid my dad's hand project was a shared adventure with surprising excursions. Instead of playing with friends my age, I wanted nothing more than to serve as his research and darkroom assistant. Flash forward fifty years and I felt both wariness and a sense of excitement about what I'd found. These boxes held everything my dad thought he had lost in the final days when our family was still intact, before the 1957 divorce.

I removed the silver masking tape from the first box and found a bundle of index cards next to a pair of long boxes filled with photographic slides tucked into glassine sleeves. Packed inside the second box I found eleven-by-fourteen-inch photographic prints of hands, intact though slightly curled. In the next box I found three-ring binders with lists of names and measurements written in a tiny precise script like some primitive Excel spreadsheet, as if I had excavated the lost language of my dad's Rosetta Stone.

By this point the dust and the heat of the garage were starting to get to me. I was about to quit for the day, but I decided to open another box, smaller and wedged behind the others. I removed a black hardcover book with no jacket and set it aside, though later I would come to know its meaning. I saw a stack of letters and pulled off the dirty gloves before pulling out stationery from Duke University dated 1952, written to my father, and at the bottom of the sheet I saw the signature of Dr. J. B. Rhine.

Under that letter I found a small envelope with a stamp and a smudged December postmark from the 1950s. On the front, the address of our home in Long Beach; on the back, the return address: 740 N. Kings Road, Los Angeles 46.

The inside card of creamy, rag-edged paper had a distinctive black-and-white woodcut design. Embossed at the top was the name Maria, at the bottom, Aldous. Inside, in a sprawling and whimsical handwriting, the message read:

> With the very best wishes for the three of you,
> From your friends the Huxleys

What was all this about? I wondered, as I began thumbing through index cards with names coded to match numbers on the mounted slides. Toward the front section of the box, I found an index card with the handwritten name Maria Huxley, Number 248, next to Number 249, Aldous Huxley.

The memory of my dad's hand project started coming back to me. What had seemed like child's play at the time now grabbed me with a grown-up compulsion. I needed to know the story behind what I could already tell was at least hundreds of photographs of hands—but mainly what I had to find out was how did my eccentric, pocket-protector-wearing engineer dad ever become Huxley's friend?

Of course, Huxley was hardly conventional, and just how unconventional I would soon learn. Even as a kid, I'd noticed that my father was not like other dads. No driving his daughter's friends to Saturday movies like Susan's policeman dad. No mixing up Sunday waffle batter like Donalee's dentist dad. On weekends my father sometimes went to the desert alone, and after workdays at Vultee Aircraft, and later at North American Aviation, my father crossed a threshold into his private world of experiments. This eventually blossomed into a seven-year-long, obsessive study of one thousand hands.

Then it all ended when he left my mother and me in 1957, shortly after his project crashed, why I never understood back then—and with it collapsed the bridge he was trying to build between science and the so-called paranormal. Eventually Dad remarried, and afterward we rarely saw each other. I moved to New York, and we saw each other even less. Then three things happened. After a decade on the East Coast I moved back to California, my stepmother died, and my elderly father came back into my life. His reappearance gave rise to my new project—chasing down the story behind the hands of Aldous Huxley.

I started with a micro recorder in my initial stab-in-the-dark attempt to understand how my father had become an exceedingly unlikely friend of the famous author of *Brave New World*. Dad agreed to meet for interviews on Friday afternoons that summer of 2001, so we set about getting reacquainted, or really getting acquainted for the first time. Except for a couple of years in the 1950s when we had been close, and like so many daughters of divorce, as a child I had known very little about my father. I guess one of my main memories was of him leaving.

There were other questions, too. Who were these names on the file cards? Surely some of these people were Dad's age by now, men and women in their eighties who had lived this story of the '50s. I wondered if I could locate some of them, or if I should even bother; and, if I did find them, could I trust their recall, or could I even trust my dad's?

I began burrowing into biographies, checking websites and obituaries, and it looked like I had mainly missed my chance for live interviews. Yet I did turn up curious biographical bits, such as finding out that hand number 247 was Gerald Heard. I learned that he was Huxley's friend and a respected BBC science commentator, as well as a bestselling author who, apart from his public persona, happened to write ghost stories on the side. I saw that one sequence of names was associated with terms like "psychic" and "trance medium," so it began to look like this collection of names might have something to do with the far side, the fringe, of science.

On the Huxley side of my investigation, I faced daunting shelves of books, mainly focused on his literary life. Hungry to get oriented, I pored over Sybille Bedford's 1974 *Aldous Huxley: A Biography*, although several other biographies had appeared since then and I rounded them up.

Soon I had filled two micro tapes with Dad's interviews, and by then I had noted a curious similarity between Aldous and my dad. Both men had undergone life-changing experiences dating back to childhood that might have sent both of them chasing after questions that eluded answers. You might say Aldous was made for the role.

Aldous Leonard Huxley was born on July 26, 1894, in Laleham, near Godalming, in the county of Surrey, in southeast England, not far from where someone whose life would intersect with Huxley's, psychiatrist Humphry Osmond, would be born a few years later.

Aldous was the youngest son in a family with a reputation for upending orthodoxy. His paternal grandfather was biologist Thomas Henry Huxley, who had mounted such a vigorous defense of Charles Darwin's theory of natural selection that T. H. Huxley became known as "Darwin's Bulldog." Growing up in the shadow of this natural science legend, the younger Huxley appeared destined for Cambridge and an eventual medical research career, but that was before three tragedies changed his course.

At age fourteen, Aldous lost his beloved mother, Julia. An admired woman who had established a school for girls, she died only two months after receiving a diagnosis of cancer. Two years later, when Aldous was sixteen, and had reached his height of six-foot-four, what seemed like pink eye was belatedly diagnosed as *keratitis punctata*, an inflammation that left him blind for a year and a half and left severe scars on his right cornea. He had to give up his studies, and during that time was privately tutored until age eighteen. Throughout much of this time he was taken under the wing of his older brother Trevenen, a student at Oxford.

In a period when I imagine day and night could seem equally dark and soon would become even darker, Aldous summoned his grandfather's tenacity by teaching himself Braille and figuring out touch-typing. During this sequestered time Aldous wrote his first novel, though the manuscript was afterward lost.

The third blow in a space of six years came about in 1914 when Trevenen, the brother who had kept an eye on the younger Aldous, fell into a deep depression over a failed love affair. Though admitted for psychological treatment, he walked away from the clinic and hung himself in the nearby woods.[2]

Aldous was motherless, he had lost most of his vision, and he had lost his favorite brother. His father had remarried in 1912 and had become more distant as he established a new family. At that time Aldous had been eighteen, with a ninety percent loss of sight in one eye and compromised vision in the other, yet he would go on to complete his education at Oxford's Balliol College, taking Firsts in English and History.

Though initial plans for a medical research career in the tradition of T. H. Huxley had been ruled out, he could look for inspiration to his great-uncle, poet and critic Matthew Arnold. He had other templates as well. His father, Leonard, was editor of the *Cornhill*, a literary magazine, and his mother before her death had been a respected educator. As he groped his way forward, these paths opened and Aldous, despite his damaged eyesight, embarked on what would become a future of writing and lecturing.

But that outcome was mainly in the future at this point. He had had a few poems published, and by the 1920s he would have four volumes of poetry to his credit. He would teach young boys and find out by sampling this career that he disliked teaching. He eventually found his way as a magazine journalist covering a broad range of subjects, in training for an unmapped future. In those early years he would write book reviews, articles on house decoration, and reviews of plays (250 plays in one year, after which he quit the job), though an advantage of theater reviews for a sight-impaired critic was, as he later said, the greater importance of dialogue compared to the lesser importance of visual effects. He would write for high- and middle-culture readers, including those who followed Condé Nast's *House and Garden*, *Vogue*, *Vanity Fair* (and later on, across the Atlantic, for *Esquire*), and had no reservation about writing for all levels of readers. While on the staff of *Vogue* he wrote art reviews, developing a technique of peering at a distance with a spyglass and up close with a magnifying glass. For a time he even rewrote advertisements for the magazine, coming to appreciate the difficulty of getting the wording for a product just right.³ Biographer Ronald Clark thinks this and the mixed-magazine discipline brought about skills that would later enable Aldous to write about abstruse subjects like mystical experience in a way people could comprehend.

Over the span of the next fifty years he would write as many books—some essay collections, some widely read novels with a few that became the bases for films, and others nonfiction books with more limited appeal because of their narrower subjects. Like Picasso drawing on a napkin, anything he wrote would sell; yet as a young man, and again in the war-disrupted middle of his life, he still had to worry about making a living.

How a man with severely compromised eyesight created such an impressive body of work is likely to puzzle anyone who thinks for a moment about it. One is tempted to say that the aftermath of his eye disease forced him to develop other strengths, including a remarkable memory. Huxley listened and absorbed. Relying on his spyglass for distance and his magnifying glass for close up, he observed minutely, and in daily life moved slowly, harvesting subtle observations that most writers might dismiss in their haste. He became masterful at perceiving and conceiving in unorthodox ways.

Like his struggle to learn Braille, Aldous would turn this intense beam toward an inner experience and eventually give it full voice in 1954. That year, in *The Doors of Perception*, he would describe the mind's-eye adventure of mescalin and show midcentury readers how to seek, and find, an elusive and arguably magical world.

On this July day in 2001, my father and I sit together on two facing chairs. We are on the patio behind my grandparents' 1910 bungalow. Concord grapes, luxuriating above us in full dress green, are still weeks away from ripening into purple. I feel strangely anxious about this dad-daughter business and wonder if it's true what they say, that you can't go home again.

Today I begin our patio session by bringing out an object from one of the boxes labeled "The Hand." Dad opens a snap on the brown leatherette case and takes out a pair of dark glasses recalling men's eyewear of the 1950s, except these lenses are pierced with hundreds of holes.

"Pinhole glasses," Dad says. "Aldous wore these."

"They belonged to Huxley?" I feel an *Antiques Road Show*–flutter at the thought.

"No, but Aldous gave me these to try," Dad says. "They're similar to the ones that helped him see. Go ahead, put them on."

For my part, I wear glasses to correct slight myopia but nothing like this type, with holes seemingly punched out by a torture device studded with nails. The eyeglasses sit heavily on my face. I survey the patio to locate a point for comparison, and it looks like the white lattice will do. I focus through the pinholes, then slide them down on the bridge of my nose and peer over the top. Up down, up down. Fuzzy lattice, sharp lattice. There is a change in perception, and I can see the difference.

I remove them blinking, and Dad laughs. "Not pretty, but they work," he says, a pet phrase of someone who values function over form. "The holes concentrate light like a pinhole camera." He begins to tell me how, as a boy, he once built a pinhole camera, then he adds a curious phrase: "People left me alone in those days."

Soon I learn something new and strange about my father, something that would connect to my understanding of Aldous and his damaged eyes. When my dad was about eight he slid down a frozen chute of ice and crashed into the side of a barn, where he lay drifting in and out of consciousness until his grandfather found him. Before the accident, this boy who would later become my father was the star reader of his school; after the accident, he could barely read a string of printed words. His mother blamed the radical change on boyish belligerence and punished him accordingly, having no idea he had suffered a brain injury probably caused by a subdural hematoma.

Today we know much about neural plasticity and adaptation to a drastic contingency, and apparently that's what happened. After enduring a period of confusion and shame, when he often hid away in the barn tinkering with his pinhole camera, he began developing an aptitude for mathematics and spatial problems that turned him from one of the school's worst readers into the school's best student in geometry and math.

What might be called short-range adaptation like my dad's also fits the biography of the nearly blind Aldous, though Huxley might be reluctant to entirely agree. This is because he played down the connection between partial blindness and achievement, though he granted that loss of sight had drastically changed his life.

I know how Aldous reached this conclusion, thanks to an interview recorded in the late 1950s on the topic "Can Physical Affliction Lead One to Creative Endeavor?"[4] On this particular windy March day a group of journalists drove up LA's Beachwood Canyon to interview Aldous in the home where he had lived for a few years by then. It was situated on a ridge, overlooking a canyon, on Deronda Drive.

As I listened to the tape in the Department of Special Collections in UCLA's Charles E. Young Research Library, I pictured Huxley in his early sixties, greeting the small party of journalists standing at his front door. One of them later described the thin, six-foot-four Huxley as reminding him of a coiled spring—this literary lion with a large head, whose childhood nickname had been Ogie, for ogre.[5] Aldous had grown into himself, with a finely shaped forehead and brow, and he gazed intensely from his better eye while at the same time focusing slightly above you.

Sitting in his favorite chair in the light-filled living room, Aldous responded to the interviewers' questions in a sonorous voice, with diction as if from a previous century. Listening to the tape and his cadence, marked by rising tones in a word such as "ac-tiv-i-ties," I fall into the pattern of how he used certain drawn-out vowels and syllables. Later on I learned it was a great tribute if, after hearing your story, he responded with a musical and almost flute-like glee with his "How absolutely in-cred-ible!"[6]

As the interview got underway, one of the journalists asked Huxley if early physical affliction made a difference in his literary achievement. At first Huxley seemed to waffle, to direct the interview away from such a simple conclusion. For one thing, he said, a year and a half of not seeing had reinforced certain tendencies, such as his inclination toward contemplation. But he did admit to a kind of adaptation. He had developed compensating techniques, techniques he had used ever since.

"Looking back," Huxley said in the interview, "I am amazed at the amount of reading I did with a small, powerful magnifying glass." It carried him through Balliol College; in the depths of blindness he had mastered the typewriter. "I even wrote an entire novel," he said, "which I never read because I couldn't read what I'd written."

The most important outcome was not creative achievement. Close friends, including six from his Eton class of eleven, were killed in the 1914–1918 war, and his affliction ("I no doubt owe my life to it," he said) kept him out of harm's way.

This was not what interviewers expected. They were looking for behavioral explanations. Aldous granted that his compromised eyesight was a kind of bridge to creative endeavor, because blindness foreclosed his chance of becoming a doctor. And so, he said, "I took to writing instead."

After poring over documents in the Huxley Collection, I kept reading my way through secondary sources, trying to understand how circumstances brought my father and Aldous to the same place on a historical timeline. Until his injury, Aldous had been on track for a patrician education and achieved this with private tutors before completion at Oxford; my father, raised on a farm, had earned his way into a mechanical engineering profession with two years of college and advancement in wartime. They were a generation apart, my father born during the Great War when Aldous was already of military age.

That first global conflict began in 1914 while Huxley was a university student who was eager, though with his affliction entirely unrealistic, about joining friends departing for the front lines. After being rejected as unfit, he joined the ranks of military-age men who had been left behind, including conscientious objectors, all of whom were expected to support the defense effort locally after the nation's soldiers marched off to war. For some who stayed home, this meant assisting with food production to reduce pressure on provisions needed to feed the troops.

Garsington Manor, not far from Oxford, was the country estate of Lady Ottoline and Philip Morrell. Part of the structure dated back to Chaucer's time. Offering young people a way to fulfill alternative service amounted to pacifism in the form of patriotism, and the Morrell's made this possible at Garsington. The manor provided a place for both young men and women, largely of the privileged class, to support the war effort by producing food and fodder in its orchards and fields.

During the years of 1916 and 1917, Huxley became a frequent on-and-off resident at this expansive country manor with its stone construction, warren-like rooms, and forecourt bordered by stately trees. Aldous toiled beside other well-bred young people in the Morrell's hayfield, where his tall frame towered over the others, and where he later came to realize that he preferred wielding a hoe to what came next in his career, which was teaching rambunctious young boys.

A set of pictures in the National Gallery in London shows him working in a dark jacket and tan jodhpurs. His fragile health and poor vision left him with the less arduous tasks like pruning branches, and I see him in another photo carrying an armful. He came to appreciate tending plants and pruning ornamental and fruit-bearing trees, an undertaking he would relish again a quarter-century later in a starkly different landscape, thousands of miles and an ocean away.

The manor house served as more than a baronial kitchen garden and a CO (conscientious objector) refuge because Lady Ottoline ran Garsington like a Bloomsbury-style salon. Parlor life at the Morrell's country house revolved around conversation, with topics ranging from free love to literature, occasionally circling back to poke fun at spiritualism, pit atheism against faith.

In such a free-wheeling yet caustic atmosphere—of which Frieda Lawrence once said of some, but not all, who gathered there, "There was no flow of the milk of human kindness in that group . . . not even a trickle"[7]—conventions were cast aside, and a young man or young woman could test their ideas and absorb or challenge the opinions of others. Aldous would encounter frequent visitor Bertrand Russell, the lover of Lady Ottoline, whose husband, Philip, didn't object to his wife's affair with the noted philosopher and logician because Philip was busy with affairs of his own.[8]

Avoidance of the clichéd and predictable in conversation extended to fashion and décor, and the larger-than-life Lady Ottoline had a bold flair for astonishment. Rooms were painted in a splash of colors—scarlet, emerald, a hallway said to be tinted the pale pinkish grey of sunrise—and her striking gowns deviated from current styles by design.[9]

Into the mix came creative types working in various genres, reflecting various sexual arrangements, some bisexual like Ottoline. Poet T. S. Eliot spent time at Garsington, as did the as-yet unpublished D. H. Lawrence, who met Aldous once and later became a close friend. Both young writers mined Garsington's quirks and contradictions and its colorful human parade to populate their early fiction. Lawrence (who once said, "There is only one Ottoline. And she has moved one's imagination") would seize upon the unattractive qualities of the redheaded, long-faced, loud-voiced Lady Ottoline to create the unlikeable character Hermione Roddice in his 1920 novel *Women in Love.* Ottoline would regard this as a betrayal.[10]

A year after publication of Lawrence's book, Aldous would complete his own Garsington-inspired novel. Ottoline would not feel as deeply betrayed by Huxley as she had by Lawrence, but she was dismayed to find herself portrayed in a novel of harsh social satire that soon became Aldous Huxley's hallmark.

The Garsington years gave Aldous a set of new tools, if not a new start, after so much early tragedy in his life. Here he learned about modern art, cultivated a nimble willingness for experimentation, was on his way to living with disregard for convention, and probably adopted the turn of mind that would allow him to stitch any idea into an essay. This and more stemmed from showing up at the Morrell's country estate for noncombatant service. In appreciation, Aldous would later say of Lady Ottoline, "She gave me a complete mental reorientation."[11]

Pivotal events often attach to certain objects, and although I have no proof, this theory might account for an object, a bit of Huxleyana, that my parents kept and that ended up as a souvenir inside one of the five boxes I found in our garage. The biography of Lady Ottoline Morrell by Miranda Seymour includes several pages of photographs, including one of Garsington Manor. In this photograph the grand country house appears on a slight rise, flanked by rows of stately yew hedges.

This composition reminds me of the woodcut on the Huxleys' personal note card, which was also, I later learn, the design of their bookplate. The woodcut portrays an imaginary library overlooking a mysterious landscape, the inside opening onto the outside world. In the foreground sits a desk with a nude statue, and through the window appears a border of trees leading the eye upward toward a stately house on a hill.

Whether the note card was commissioned or selected ready-made and personalized with the names Maria and Aldous, I do not know, but to me it evokes Huxley's portal to independent adulthood and the setting of his courtship of Maria. And I now know that the whimsical handwriting I first saw on the note card has something to do with how a young refugee who almost killed herself survived her years at Garsington to become Mrs. Aldous Huxley.

It was a courtship arising from wartime chaos. A few refugees, not picked randomly from among the masses of the displaced but a few individuals from well-connected families, also resided at Ottoline and Philip Morrell's country estate. One was a dark-haired girl named Maria Nys, whose family had fled their home in Belgium in 1914, escaping just ahead of advancing German troops.

The family's exile began in late summer of that year, after Madame Nys and her daughters left Bellem in northern Belgium, reached Ostende, and from there crossed the English Channel. Following referrals from friends, Madame Nys sought a suitable refuge for herself and her girls, but unable to find accommodations that would allow them to remain together it was decided that Maria would stay at Garsington.

The daughter of an industrialist and raised in a life of affluent ease, Maria gave other houseguests the impression of being a willful and spoiled young woman, perhaps a defense for someone who found herself living with strangers and had no financial resources of her own. The plump lively girl, with glossy dark hair and expressive blue-green eyes, with a small full mouth and a small pointed chin (as her sister in-law-would later describe her) would often turn shy in the presence of an ensemble of older writers and artists who shared a language and culture unlike her own.

Maria remade herself by fashioning a persona one contemporary described as "fastidious elegance with a touch of fantasy." In an unconventional household like Garsington, where codes were fluid and misunderstandings frequently arose concerning affairs of the heart, Maria fell into a state of longing for her benefactress, Lady Ottoline, and, apparently rebuffed in the spring of 1915, the distraught sixteen-year-old swallowed chloride in an attempt to take her own life.[12] Discovered in time and placed under a doctor's care, it was from this low point that she began adjusting to her refugee life by tapping inner resources no one knew she had. This was the girl who grew into the woman with the whimsical handwriting.

A few months later, in December of 1915, Lady Ottoline invited a shy, twenty-year-old Oxford student of English literature to visit Garsington. After their first meeting over lunch, Ottoline wrote her first impression of Aldous Huxley in her diary, describing him as shy and aloof, but perhaps this was understandable given his brilliant family, with his elder brother, Julian, a respected biologist, his father an admired magazine editor, his maternal grandfather a noted man of letters, and his paternal grandfather Darwin's knight and defender.

After that first luncheon meeting, and over the next two years on weekends and during summers and over university holidays, Aldous often stayed at Garsington, where his place was always set at Ottoline's feast. He was well liked by others, a tall young man with abundant dark hair and alluringly strange eyes, someone you could almost call handsome, whose visible frailty placed limits on his physical activities but not on the scope of his mind.

Aldous began to focus his attention on Maria in the summer of 1916, and this set a tentative courtship in motion. At first showing little interest, she gradually responded in ways one might imagine, over meals in a formal dining room or passing by each other in hallways, until affection turned into love.

In one photograph of Aldous taken in front of a fireplace in the Red Drawing Room at Garsington he is wearing glasses, with his thick, dark hair framing his fine features, his long legs crossed as he sits in a wingback chair. He looks away from the camera with his face half hidden, holding a book unnaturally close.

Maria looks directly at the camera in a photo of her in the same chair. Both were taken a couple of years after they left, during a return visit. Maria was once a plump and unhappy girl when she arrived at the Morrell's country house, but now she is a slender, stylish, and lovely young woman. She wears a draped skirt with a delicately pleated white blouse, her dark hair fashioned with a touch of whimsy into a pair of chignons hugging her ears. Her hands convey action and repose, one touching a desk slightly, as if poised in mid-task. The fingers of the other hand rest softly on the upholstered arm. Unlike Huxley's half-hidden face, her direct expression is revealing of their future together. She will be the one welcoming visitors and filtering their requests, the first to rise to the needs of family and friends who will enter Aldous Huxley's close-focused world.

I think it is fair to say that, compared to young men of his advantaged social class, Aldous needed a dedicated partner to help him fulfill his potential. In late 1916 he proposed marriage on the lawn at Garsington, even though Aldous had no employment, it was still wartime, and Maria's family was scattered. The couple vowed to wait for each other, not knowing the delay would last more than two years.

Soon Maria's mother collected her four girls and moved the family to Florence while Aldous remained behind. He accepted a position as schoolmaster at Eton, where one of his students, Eric Arthur Blair, would later write under the pen name George Orwell and fashion a dystopian novel often compared to one by his teacher; but at this point on the timeline *Brave New World* and *1984* were years away. While a schoolmaster at Eton, Huxley realized that he did not enjoy teaching young boys; moreover, it paid too little to support his future family. He had a superior education, but with his father remarried and his father's second family needing their own posh educations, Aldous had to be resourceful about working out his own prospects. Soon he was rescued from the job of schoolmaster by an offer of an editorial position at *The Athenaeum,* a respected London-based literary magazine, and once he began moonlighting reviews a magazine journalism career unfolded.

On the eleventh day of the eleventh month of 1918 the war ended, leaving millions dead in its wake and a lingering horror of the carnage of global warfare. Shell-shocked normalcy returned to England and Belgium, and, after an engagement of two years and three months, Aldous and Maria were married in July of 1919 at her grandparents' home in Belgium. She was twenty, he twenty-five. They began their life together in a Hampstead flat twenty minutes by train ride from his work near Charing Cross, sometimes called the "exact centre" of London. Nine months after the wedding, in April of 1920, Maria gave birth to a boy they named Matthew, who would be their only child.

After their wartime courtship at Garsington, with its captive-audience amours and Bloomsbury-style progressive ways, it seems inevitable that in coming years the young couple would reject many of the conventions of matrimony. They would travel the world together in a marriage bonded by love and respect and a shared effort to support his work, yet it would be a marriage loosely tethered by creative consent that a later generation would call an open marriage.

Garsington memories provided a literary dowry to furnish Huxley's successful 1921 novel *Crome Yellow*. Fleeting impressions and whispered assignations, colorful rooms and flamboyant costumes, conversations during supper and overheard in hallways and even on the roof all found their way from his notebook and memory into paragraphs and scenes and chapters. After *Crome*'s critical acclaim, Huxley gained a reputation for wielding a sharp pen, flitting into parlors like a modernist gadfly and skewering members of an ossified social class.

The cloak of satire also gave Aldous a chance to bring up other issues and ideas in the dialog between characters. Even in *Crome Yellow*, hints appear that Huxley had one eye on the fringe of science. Occasionally his characters banter about the realm of the occult, and certainly spiritualism was in vogue after the Great War when sons and husbands died in battle and families were left longing for contact. Some turned to séances presided over by an individual called a trance medium.

Apart from portraying characters that occasionally express these ideas, I see no evidence yet that Aldous had developed an open mind about the possibility of an afterlife, as he would eventually. Much of his longer writing through the 1950s reflects a sense of horror toward death,[13] understandable in the aftermath of losing his mother to cancer and his brother to suicide and former classmates to trench warfare.

The grandson of Thomas Henry Huxley, who coined the word agnostic, might have found the notion of survival after death simultaneously preposterous and tantalizing. Grandfather T. H. wrote, "I neither deny nor affirm the immortality of man. I see no reason for believing it, but on the other hand, I have no means of disproving it."[14] Years later, in the summer of 1961, in an interview with journalist John Chandos, Huxley would say that in his youth before the war years he had read classic descriptions of the mystical life by Jacob Boehme and William Blake with what he called a mix of skepticism and "fascinated interest."[15]

Conventional religion was easy to mock for hypocrisy but not so easy to mock as personal experience. One of Huxley's evolving fiction techniques, which some have called essays masked as novels, was creating characters with opposing views he wanted to explore. He had listened to Lady Ottoline and Bertrand Russell arguing about faith versus atheism in debates where Russell always won, but Ottoline walked away with her deepest convictions intact. Hers were not dogmatic convictions but eclectic, as suggested by her passion for the contemplative manual by Thomas à Kempis, *Imitation of Christ*, as well her interest in professor William James's *Varieties of Religious Experience*.[16]

Scraps of dialog from his recent schoolmaster days at Eton

crept into Aldous's first novel. His students had become interested in Theosophy, a metaphysical philosophy or quasi-religion based on Madame Blavatsky's vision of other dimensions, an amalgam of so-called ancient wisdom spun into cosmic laws and astral planes. Apparently when his students brought up Theosophy, Aldous countered with facts and by pointing out inconsistencies, but he noted proof by scientific method did not apply to personal religious experience. When it came to the core of the world's great religions, Aldous taught his Eton students about tolerance, that each religion contained a seed of truth.[17]

CHAPTER 2
SEEKERS OF PEACE

French novelist Marcel Proust, another observer of the leisure class, called Aldous Huxley one of the most promising young writers of the day. This comment came before Proust's death in 1922 left the last of his seven-volume novel *Remembrance of Lost Time* unfinished. Huxley would be similarly prolific, and after achieving an international literary reputation with *Crome Yellow* he would go on to author seventeen books, many of them collected short fiction and essays, in the decade between 1922 and 1932.

By 1923 Aldous was under contract to publisher Chatto & Windus and free to write wherever he pleased. The Huxleys spent time in Belgium and Italy, where Maria ordered, factory-direct, a red Bugatti roadster with pale grey seats and had it modified to suit her husband's long legs. A highly skilled driver and an amateur mechanic herself, Maria is said to have had an uncanny intuition when something was amiss under the hood. The Bugatti was a guaranteed source of enjoyment, prompting Aldous in a 1931 essay to write, "Speed, it seems to me, provides the one genuinely modern pleasure."[1]

The success of his early novels gave Huxley more freedom to write essays of his choosing, unlike his journalism writing-for-hire days, and the essay side of his career expanded beyond reviews to include travel writing. In 1926 the couple set out on an around-the-world trip from India, through Southeast Asia and Indonesia, across the Pacific to San Francisco, then south by train for Huxley's first taste of brash, boom-era Los Angeles, which this first time around he clearly didn't like. Among the recipients of his letters was Mary Hutchinson, with whom he had an affair during the '20s, one of the progressive triangular sort (some suggesting that Mary preferred Maria).[2] Another of Aldous's paramours in the early '20s, briefly though intensely, was Nancy Cunard, daughter of the heir to the shipping line. Maria is said to have had a hand in courtships, sending flowers and making arrangements.[3] Descriptions of both Hutchinson and Cunard found their way into novels written on a contract deadline, specifically his 1923 *Antic Hay*.

In 1930 they purchased a home in the south of France at Sanary-sur-Mer.

Huxley's biographer Sybille Bedford, his then-neighbor in this seaside town, describes Aldous at home on a typical day as wearing sandals and khaki shorts with a cotton shirt. He worked in a study with a red tile floor, and wrote his next two novels, *Brave New World* and *Eyeless in Gaza*, on a small typewriter, metal keys plunking away. He also painted in an upstairs studio, and sometimes the sitter was a child.[4]

In the first of the two novels, Huxley created a dictator-controlled society where a Pavlovian reward and punishment, backed by pacifying euphoric drugs, kept artificially imposed social castes in line. *Brave New World* would become his most enduring work, though it was not widely popular in its publication year of 1932 when perhaps the plot was too short a leap from newspaper headlines. A worldwide economic depression with consequent social distress had fueled support for the authoritarian fascism espoused by Italian Prime Minister Benito Mussolini, and his success gave Hitler further impetus to use strong-arm tactics. Huxley's vision plunged the reader into a techno-fascist society where individuals had long since relinquished any control over their own fate.

The second novel Huxley wrote in Sanary was *Eyeless in Gaza*. Its title, from John Milton's poem *Samson Agonistes*, alludes to the cruelly blinded biblical figure of Samson, an apt image for what is said to be Huxley's most semi-autobiographical novel. Begun in the early 1930s and published in 1936, and unlike his novels with aloof characters lifted from a Garsington house party or *Brave New World* with its chilling social landscape, *Eyeless in Gaza* opens up Huxley's own wounds. He exposes the pain of a mother's death and portrays the alienation of a stepmother's household. The character Brian Foxe commits suicide.

One of the other Sanary residents was Charlotte Wolff, a psychotherapist who had fled Nazi Berlin and now made her living by consulting through hand analysis. Aldous and Maria became interested in Wolff's approach to understanding character through the human hand, one of the typologies that would come to fascinate Aldous.[5] Maria began collecting handprints of people, many of them famous, that she and Aldous came across during their travels.

The hand theme had first appeared a decade earlier in Huxley's 1925 novel *Those Barren Leaves*, where early on the author describes a haunting image of hands backlit by firelight. A scene rounds out the novel with a main character reflecting that if one looks long enough, say at one's hands, it may be possible to probe to the bottom of all the mysteries.

By 1936 Huxley had begun exploring meditation and hypnosis. That July he wrote to a friend, "I think a great deal can be done to modify oneself. It's a question of using the proper techniques: meditation, Mental Prayer, whatever one likes to call them."[6]

Cultivation of the inner world, however, began to take a backseat to unfolding world events. Chancellor Adolf Hitler's brand of fascism, passed off as a form of socialism, was becoming increasingly aggressive as he eyed the borderlands and envisioned colonies of pure blooded Germans populating an expanded *Reich* or empire.

Having lost a million sons in the European conflict of 1914–1918, the British were wary of involvement. Huxley had lost friends in what was then called the Great War, "the war to end all wars," and supporting this hope Aldous turned idealism into advocacy by giving his backing to the Peace Pledge Union headed by the charismatic Anglican priest Reverend H. R. L. Sheppard. Huxley embraced this effort in tandem with Gerald Heard, whom he had first met in London in 1929. The respected BBC commentator had an encyclopedic mind much like Huxley's and would be largely credited with leading Huxley toward spiritual exploration and experimentation. It began with the practical idealism of their participation in the Peace Pledge Union.[7]

Heard, though known for covering a wide range of anthropological and scientific topics, quietly pursued more controversial interests as well. These included active membership in the London Society for Psychical Research, which had been founded in 1882 and was most often associated with studies of haunting phenomena, clairvoyance, and so-called out-of-body experiences. Its mission then and now is to support research into human experiences that challenge contemporary scientific models.

Huxley and Heard's shared focus in 1937 was not about matters on the fringe of science but about the saber-rattling threat of war. Convinced that peace must be defended as an ethical imperative, the two planned a lecture tour to help persuade Americans to remain neutral, remain on the sidelines, and thereby increase the chance of averting a global war.

With this goal on April 7, 1937, Aldous and Maria boarded the two-year-old luxury liner SS *Normandie* with their son Matthew, Gerald Heard, and his friend Christopher Wood. The transatlantic liner with its sleek furnishings, sweeping murals, and Lalique chandeliers was a floating showcase of Art Deco design. On departure day, the Huxleys may not have sipped bon voyage champagne from crystal stemware, because they were traveling third class, though when French Line officials saw their name on the passenger list they insisted on giving Huxley's party a complimentary upgrade.[8] Much of the civilized world was in the process of a downgrade as conditions in Europe continued to deteriorate.

That year of 1937 may seem like a premature date for talking about United States involvement in what became the Second World War, but I learned that war preparation was already underway at that point on our side of the Atlantic. I heard about this during one of the first interviews with my father on a day I had expected to talk about Aldous; but that day Dad wanted to talk about war.

I had barely tested the micro-recorder when a buzz saw began ripping away in the neighborhood somewhere near my grandparents' bungalow. When the ear-numbing sound subsided, Dad and I began the day's interview.

By now I knew a few details about his role in the war effort; I knew he had written a book called *Aircraft Lofting and Template Design,* which was used in National Defense classes to train draftsmen for production of the BT-13 trainer-fighter aircraft, a plane considered crucial in winning the war.

"It took four years to develop the first BT-13 and in the end we built thousands. They were already working on the design in 1937 when I signed on at Vultee Aircraft," Dad said. "Someone knew this war was coming."

Until the late 1930s, aircraft manufacturers like Vultee, in Downey, California, had only produced a few airplanes each year. Made of wood and aluminum, with parts largely shaped by hand, the craft were essentially one-offs. Vultee tried a faster production line with war, or the war materials business, likely on the way, only to find that the first versions had problems with the simplified wing design. To solve this, a consultant was brought in from the shipbuilding industry, a master designer with expertise in handling the curvature of keels and hulls.

As my father tells it, the consultant wanted to work alone to protect his trade secrets, but Vultee demanded that he collaborate with one of their own. This assignment was given to my father, who had earned a reputation for coming up with unusual design solutions. He did this by saturating his mind with the problem. The process then somehow continued while he slept at night, and often a solution came to him in the morning. My father and the master shipbuilder worked on blueprints drawn in three dimensions (lofting entails the transfer of lines to a full-sized plan), and through their collaboration developed a drafting technique that would help move the BT-13 into mass production.

By this point in the 1937 timeline, Aldous and Gerald had arrived in New York, but before the Peace Pledge lecture tour began, Aldous had a task to accomplish for the sake of an old friend. Starting from New York, where newsreels documented the Huxleys' arrival on the SS *Normandie*, the family and Gerald Heard set out on a five-week road trip, first purchasing a new brown Ford before motoring southwesterly with Maria behind the wheel. Once past the major cities she must have relished the feeling of freedom as ribbons of road disappeared under her hood, even if driving a Ford was no match for her scarlet Bugatti.

Their first destination was the New Mexico ranch of the widowed Frieda Lawrence. D. H. Lawrence had died seven years before in the south of France with his friends the Huxleys by his side. Now settled in for a few weeks in the New Mexico highlands, freshened with summer breezes, Huxley edited D. H. Lawrence's papers and recalled the days when the two hopeful writers first met at Garsington under Lady Ottoline's wing.

By summer's end the Lawrence editing project was done, and weather was about to turn conducive to a lecture tour inside auditoriums and assembly halls. Continuing their road trip, the party traveled through Colorado to deliver Matthew to a prep school. Its

US location was a hedge against British military conscription, which, apart from constant worry over Matthew's safety, would have been anathema to his pacifist father.

They motored on to Hollywood, where a newly forged friendship with screenwriter Anita Loos (they had met in person that past spring in New York, though they had corresponded before that) would later bring scriptwriting opportunities. Aldous and Gerald launched their cross-country lecture tour at the Philharmonic Auditorium in Los Angeles, a venue that would play a future role in this unfolding story.

In November of 1937, while Huxley and Heard were on the road fulfilling their Peace Pledge commitment, Adolf Hitler held a small conference in the Reich Chancellery in Berlin. Demanding a vow of secrecy from six men, including Luftwaffe Commander Hermann Göring, Hitler announced that under a policy of *lebensraum,* or territorial expansion to gain living space he felt was deserved by a superior race, he would act to take land beyond German borders by force.

Huxley and Heard, unaware of the futility of their pacifist cause, continued the Peace Pledge lecture tour until Gerald broke his arm in Iowa and needed time to recover. For a while Aldous kept up their schedule, though Gerald's BBC background made him by far the better public speaker of the two. Now Aldous was on the lectern alone, and it would be years before he learned to enjoy giving a lecture. They did not resume the tour jointly, and after Gerald's arm mended he traveled to Durham, North Carolina, with the intention of taking up a position he had been offered as chair of Historical Anthropology at Duke University.

At this point in the timeline, a new story arc begins to take shape. For the past five years, Gerald had served on the governing council of the London Society for Psychical Research. Perhaps not coincidentally, his academic appointment at Duke would take him to the home of Joseph B. Rhine's parapsychology laboratory. The laboratory had been funded through the largesse of one of Dr. Rhine's first research subjects, an internationally proclaimed psychic and trance medium named Eileen Garrett, who herself had been bankrolled by a grateful client. Within a decade, Eileen would become a member of Huxley's experimental circle of friends.

Professor Rhine's parapsychology lab had begun conducting experiments with telepathy and precognition as early as the 1920s. Rhine's book *Extra-Sensory Perception* arrived in bookstores in 1932, the same year as Huxley's novel *Brave New World*. Initially called ESP, within a few years the preferred term became *psi*, referring to a letter of the Greek alphabet and serving as a more neutral term by avoiding the connotation of an external force at play.

Narrowing his research to what could be quantified, Rhine's book was based on ninety thousand laboratory trials using sets of Zener cards. A set consisted of twenty-five cards, each printed with one of five symbols: a cross, wavy lines, a star, a circle, or a square. The test using these cards tracks whether a specific visual image can be transmitted between two subjects who are isolated in separate rooms. Results from the rigorous but often tiresome Zener tests are scored and analyzed in an effort to substantiate evidence of psychic activity.

By contrast, the qualitative approach of describing anecdotal, spontaneous, one-off cases of telepathy and precognition, reported by individuals who claim remote knowledge of actions usually affecting a loved one, is not subject to the controls and replication of the laboratory setting. In the framework of statistical verification, such accounts of anomalous phenomena simply do not count.

After the lecture tour ended, the Huxleys made another southwest swing for a visit with D. H. Lawrence's widow, Frieda, before returning to Hollywood. Aldous and Maria had anticipated returning to Europe in 1938, but in March of that year Nazi forces occupied Austria. A certain percentage of the population warmly welcomed the Führer, and Hitler's *lebensraum* push for more living space rolled on. Six months later, leaders from Italy, Great Britain, and France signed the Munich Agreement, permitting Germany to annex the German-speaking territory of the Czechoslovak Republic. This was the turning point, Matthew Huxley told biographer David King Dunaway, when the young man realized that he and his parents would not be going home. Concerned, too, about covering living expenses, Aldous and Maria decided it would be wiser to stay in the States.

In the United States, saber rattling didn't hurt the film business. As it had done for years during the protracted economic depression, entertainment provided the escape of virtual travel to times and places real or imagined. Huxley had been encouraged to write for the film industry by rare-book dealer Jake Zeitlin, who would form a decades-long association with Aldous; and he had also been encouraged by Anita Loos, one of Hollywood's first female screenwriters, who pointed out that the studios were eager to hire men of his caliber. The movie-going public couldn't get enough of British wit, charm, and erudite dialogue, she told him.[9]

Once back in Southern California, Aldous dipped into the stream of spirituality and studied for a time with Swami Prabhavananda, the guru of the Vedanta temple in Hollywood. Though in many ways Westernized and a smoker, the Swami followed classical Hindu devotional practice, and under him Aldous took up this style of meditation.[10]

Huxley was already familiar with many features of the Indian philosophical tradition. His *Brave New World* society revolved around a pleasure-inducing drug called *soma*, a term borrowed from ancient Vedic scriptures in which it referred to a food of the gods. Meditation suited Huxley's inward inclination, but eventually he would chafe against the rigid rituals of the guru system. Around the same time, Gerald arranged the introduction of another guru, or an anti-guru, named Jiddu Krishnamurti, whose less structured spirituality would turn out to be a better fit for Aldous.[11]

Jiddu Krishnamurti had been a Brahmin-caste boy of fourteen

living in Madras, India, when in 1909 Annie Besant of the Theosophical Society, a successor to founder Madame Blavatsky, chose him to be groomed as the next World Teacher. Initially the Theosophical Society had two centers in the States, in Chicago and in Hollywood. In 1922 Krishnamurti, then twenty-seven and leader of the religious organization, spoke at the Hollywood Bowl to an audience of more than sixteen thousand. That same year, Besant bought acreage in Ojai, a two-hour drive north of Los Angeles, and in 1924 Krishnamurti began giving talks in the rustic outdoor setting of Ojai's Oak Grove.

In 1929, with the blessing of Annie Besant, Krishnamurti broke away from the Theosophical Society, though he continued to live and lecture and receive visitors in Ojai. Krishnaji, as his followers and friends called him, expounded a philosophy of discrimination between the real and the unreal and rejected devotional religion's rituals.

Under this iconoclastic approach, where no-meditation is meditation and vice versa, Krishnaji took long walks sometimes accompanied by others, in which both the walk and the surrounding nature served as the meditation. In his view, answers arose from silence if you knew how to listen. When someone brought him a problem for consultation it seemed to untangle wordlessly even before being verbally posed to Krishnaji, and when he did respond he did so obliquely, so that between the lines answers might flow.

Huxley would find this approach so unusual and arresting that he encouraged Krishnamurti to publish his daily notes in what later became the multi-volume *Commentaries for Living*. The inside dust jacket of any volume suggests that the reader "start by reading a commentary for just one minute," thereby opening a door.

Aldous, who was hospitalized in 1938 with a severe case of bronchitis followed by a slow recovery, sent word to Gerald that he wanted to meet this person he'd heard about, this Krishnamurti, this man who had resigned gurudom and struck out on his own. Aldous had also heard that he was a proponent of pacifism. So in May, when at last Aldous was well enough to travel, he and Maria drove with Gerald to Ojai and spent the day in the country. On their initial meeting, Krishnamurti recorded in his notes, "Of course Huxley is what's called an intellectual but I don't think merely that."[12]

That summer Aldous worked on a film script of *Madame Curie* and his financial situation improved, though he disliked the necessity of sitting in an office at MGM. In his off hours he began writing a satirical novel that would become *After Many a Summer Dies the Swan*, a story involving a Randolph Hearst-like mogul and a Marion Davies-inspired mistress. The 1941 Orson Welles film *Citizen Kane* would also be inspired by the lives of Hearst and Davies, but Huxley's main protagonist is an unlikely-for-Hollywood character named Propter, modeled after Gerald Heard, who unlike most denizens of the film capital craves self-knowledge rather than a marquee name.

When Aldous wasn't working at the studio or tapping the typewriter at work on his latest novel, he and Maria often socialized with a circle of British expats who disagreed on whether to return or remain in the States if Britain went to war. One was writer Christopher Isherwood, who had been recently introduced to Huxley. This circle of Brits overlapped with a European émigré community of artists and intellectuals, and both circles occasionally overlapped with the film community.

The Huxleys loved picnics, and biographer David King Dunaway recounts one occasion in mid-1939 when these overlapping worlds came together in a picnic gone awry. The group, which set out in several cars, included Greta Garbo, Charlie Chaplin, Paulette Goddard, the Huxleys with their son, Matthew, and Krishnamurti with colleagues from India who, being vegetarians, brought along their own cooking pots and food.

The would-be picnickers found an ideal spot, except for the posted No Trespassing sign. In no time a peace officer showed up with gun in hand, and because they were dressed so casually (or exotically, in the case of the Indian visitors) he mistook them for vagrants. Aldous explained who they were, but the officer didn't believe a word and instead told the ragtag group of deep incognito film stars and odd-looking visitors to take their paraphernalia and leave at once or he would have them all jailed.[13]

This took place around the time of Huxley's forty-fifth birthday, when he had just completed *After Many a Summer Dies the Swan*, but its publication and the prospect of royalties was months away. Making a living from writing was much dicier in a world on the brink of war, and again he began casting about for income. Within a few weeks MGM had offered him a project, adapting Jane Austen's nineteenth-century novel *Pride and Prejudice*, and it came with a whopping salary.

That should have been good news, except it came against the backdrop of tragedy unfolding across the Atlantic. On September 1, 1939, Germany invaded Poland. The next day German U-boats sank the British-flag Cunard passenger ship *Athenia*, which carried 1,103 passengers, killing 118. On September 3, Britain and France declared war on Germany. Perhaps the Peace Pledge lectures contributed to what came next because the United States claimed neutrality on September 5.

In light of all this, MGM's weekly salary offer of $2,500 seemed shocking. Aldous was prepared on principle to refuse the job, until the Huxleys' friend Anita Loos cut through the fog of indecision with a practical line of dialogue: "Why can't you accept the money and send it to England?"[14]

CHAPTER 3

JACOB'S HANDS

Rationing began that January in Britain as Nazi forces began marching through the continent. Maria's native Belgium, where family members still lived, surrendered in May. Bombs fell on Paris in June. The Battle of Britain began in July, followed by the Blitz. Germany ended the year with a massive air raid on London.

The scope of the war continued to expand and battles took place in the Atlantic and Mediterranean, including one at sea in the spring of 1941 involving a young doctor named Humphry Osmond. Later Osmond would become a central figure in Huxley's experimental circle of friends, but at this point in the timeline he was in the North Atlantic serving in the British Navy, where his ship escorted convoys from Iceland and Norway to Casablanca and Gibraltar, wherever German U-boats stalked their prey.[1]

As a medical officer deployed with 126 others on a 312-foot V-class destroyer, Osmond was dedicated to keeping his men alive. In one desperate convoy battle, two British ships were sunk, and he could only watch helplessly from the bridge. Despite rescue attempts, hundreds of British sailors drowned or died from their injuries or were killed in shark attacks, and he would be tormented for years by the memory of so many men lost at sea on that one terrible day.

In December of 2007, I stood newly arrived from the airport in an unfamiliar living room decorated for the holidays. One of the first things I saw was a framed marine oil painting hanging on the wall. "That was my father's ship during the war," said Humphry Osmond's daughter Fee, short for Euphemia, who spoke with a soft British accent although she had lived much of her life in the States. The marine painting, portraying action on the high seas, hung on the wall of her Wisconsin home where I was her houseguest that icy winter weekend. Her father had died a few years earlier, and she had invited me to help her sort through his photographs, books, and memorabilia.

The air was fragrant with the scent of cinnamon rolls, and beyond the plate glass window the first snowfall of winter had formed a peak on the bird feeder, a far cry from my green lawn in Southern California.

"I remember seeing a mention in one of your father's letters of how he read Huxley's books during his off-duty time aboard this ship," I told Fee.

Osmond had singled out as a favorite Huxley's 1932 anthology called *Texts and Pretexts,* its subtitle *An Anthology with Commentaries.* I had not seen the book until then, but there was Osmond's personal copy sitting in a nearby shelf so I opened it. The first line of Huxley's text read "An anthology compiled in mid-slump?" It seemed the mid-slump referred to its publication date of 1932, in the middle of a worldwide depression. Between these covers Huxley had assembled essays and poems worth reflecting on, work by William Blake and Emily Brontë, Walt Whitman, and others. I could see how this might help someone get a handle on a personal case of "mid-slump." A telling quote from one of Huxley's commentaries on page five is, "Experience is not what happens to a man; it is what a man does with what happens to him." I'm sure such words offered perspective if not consolation after what Osmond had witnessed at sea.

A decade later and a few months after they had finally met, Humphry Osmond would tell Aldous how much this book had meant to him, and how he had never imagined meeting Huxley, much less collaborating and calling him friend. Nor would Osmond imagine in the early days of knowing Aldous how their collaboration would have a big effect, for better and for worse, on his family, just as my father's association with Huxley would make its mark on mine.

Roughly nine months after Osmond's convoy battle, a formation of Japanese planes lifted, one by one, from the deck of an aircraft carrier secretly positioned in the Pacific, their destination Oahu, Hawaii. Reaching the target at 8 a.m. on December 7, 1941, the Japanese airmen bombarded the US Navy fleet at Pearl Harbor, destroying or damaging eight ships and knocking out combat planes, killing over 2,400 Americans.

The next day, President Franklin D. Roosevelt declared war against Japan.

With its Pacific fleet demolished and manufacturing in the relative limbo of two years of neutrality, President Roosevelt said in an address to the nation, "Never before have we had so little time in which to do so much."[2]

To ramp up the war machine in support of the US War Department, major manufacturers sent representatives to emergency multiday meetings on both coasts. One of these think tanks was held in Burbank, California, at the offices of the Walt Disney Studios. Because of its sensitive location near the Lockheed airfield, the buildings had been appropriated by the military within hours of the Pearl Harbor air raid.

My father, age twenty-five (the same age as Humphry Osmond, whom he had not yet met), attended the Burbank meeting as the representative from Vultee Aircraft, where he shared a room with an engineer from General Motors. Recalling the meeting and its purpose, my father said, "They had never done a production line of airplanes as they had with automobiles. That bridge from airplanes as one-offs to assembly line was like walking up to the Colorado River and saying, 'How do I get across this son of a bitch?' We were dealing with the concept of jumping across the river."

While my father became enmeshed in wartime work, and Fee's father served on a destroyer in the Atlantic, the Huxleys had already taken several steps closer to settling in as expatriates. Ever since their arrival in Southern California, they had moved between a succession of furnished homes in the Los Angeles–Hollywood–Beverly Hills area— on North Laurel Avenue, North Linden Drive, Crescent Heights near Fountain—but by end of 1939, it being evident that they would not recross the Atlantic in the foreseeable future, they rented a furnished home on Amalfi Drive in Pacific Palisades for what turned out to be three years. On this broad street lined with eucalyptus and pine and redolent of salt air, where several of their neighbors were well known in the film world, the Huxleys lived from 1939 until January of 1942.

The cusp of 1942 was another turning point for Aldous. An advocate for peace, he became despondent when the last hope of avoiding a full global conflict was shattered by Pearl Harbor. Perhaps it was a coincidence, but the Huxleys were on the move again shortly after the US declared war. One practical reason was health, his and hers. He had suffered from bronchitis, and Maria recently had been diagnosed with a mild case of tuberculosis, and both conditions called for a change from the dampness of Pacific Palisades and the smoggy air of LA.

Tethered somewhat by the studios, though no longer tied to a desk at MGM, they decided to look for a home in the desert around a two-hour drive away from the film capital. They planned to give up the large Amalfi Drive house, but because Aldous would have occasional meetings in town they would keep a small flat in Beverly Hills south of Wilshire. Now with a plan in place, they began a search for a rustic getaway in the sparsely populated expanse of the Mojave Desert.

Business as usual took a backseat in wartime. The same American manufacturing plants that had recently turned out consumer goods ranging from automobiles and tires to appliances and ladies' nylons were retooled for the defense industry.

In the aftermath of the attack on Hawaii, both men and women volunteers flooded US Army recruiting offices. Registration for the nation's first peacetime draft had been underway since it was signed into law in the autumn of 1940. The law had a provision for conscientious objectors, and other draft-age men would be exempt if their skills were needed for certain civilian jobs.

To fill those skilled jobs, never demanded in such numbers before, one of the conclusions reached at the 1942 manufacturing think tank in Burbank was that in a time of war emergency it was important to locate and harness latent talent. In an era of blueprints before computer-aided design and photocopies, many draftsmen had to be trained from scratch. Also in that era, long before the GI bill brought higher education to so many, most of the men who would fill certain categories of manufacturing jobs were high school graduates or had limited college credits. Under wartime pressure, the potential for filling a critical role had to be identified early in job recruitment.

Vultee Aircraft's mandate was to get planes into production as fast as possible, and some shortcuts were unorthodox in terms of what we would call human resources. In my father's area of expertise, the lofting or wing specifications, this required finding talented candidates. My father, certified by UCLA as a college-level wartime instructor, was tasked with designing a series of classes that would ready as many draftsmen for the aircraft plant as possible. To accomplish this he devised an aptitude workbook with a step-up sequence of geometry problems that identified those with a knack for the work. Conversely, the exercises washed out students who did not swiftly grasp the concepts. There was no time for the usual progression of college courses.

A string of coincidences led to Dad's eureka moment and the

birth of what we came to call the Hand Project. After a would-be draftsman demonstrated his potential, my father would sometimes shake hands with the successful student. My father also shook the hands of some of the unsuccessful ones, wishing them good luck with their next endeavor, which probably meant losing draft deferment status and being shipped off to boot camp. Such tactile impressions, those handshakes, seem to have accumulated subconsciously until one day he noticed that the two groups (finger length, perhaps, or palm width, or maybe palm thickness) felt different during a handshake.

"It got me wondering," he recalled. "If I measured the hands or maybe took photographs, could I quantify that difference?"

The move to the Mojave Desert, and wartime tire and gasoline restrictions, interrupted Huxley's occasional presence at the Vedanta temple, but he would find other ways to continue his quest for ordinary—and extraordinary—perception. California has a longstanding reputation as a magnet for divergent religions and occult practices, perhaps more so than any other place in the nation. The first wave swept in with the economic high tide of the 1920s that brought droves of Midwest migrants seeking their fortunes, among them evangelist Aimee Semple McPherson, who regularly preached to Los Angeles crowds of thousands in her more or less aptly named Angelus Temple.

The treasures revealed in Tutankhamen's tomb in 1922 also caused a sensation and sparked a feverish interest in ancient religions. Manly P. Hall, whose encyclopedic mind flowed nimbly across an eclectic range of arcane traditions, also began lecturing to swelling crowds. The Theosophical Society came to California in the 1920s, as did another organization claiming an ancient lineage and adopting Egyptian symbolism, the Rosicrucian Order (AMORC—the Ancient Mystical Order Rosae Crucis), which had established its headquarters in San Jose somewhat earlier, in 1915, and also had satellite groups in communities around Southern California.

If the first wave mirrored boom times, another wave of esoteric religions and occult practices in the early 1930s may have reflected the mood of the Depression. The Vedanta Society of Southern California, its roots in the ancient Hindu scriptures, established a temple in Hollywood in 1930. Manly P. Hall, who had been lecturing on such topics for a decade, founded the Philosophical Research Society in 1934 and built its headquarters in the Los Feliz district (where a street named Huxley, after Aldous, today runs perpendicular, a block away). Lesser-known groups devoted to tapping the power of esoteric knowledge offered their own trappings and secret practices. One such group built a sanctuary in Laguna Beach using rubble from the 1933 Long Beach earthquake and named their church for St. Francis of Assisi.

Huxley had noticed the profusion of religious offshoots on his first visit to Los Angeles in May of 1926, during a brief stop on his journey around the world, and he wrote about it in the travel diary published as *Jesting Pilate*. Not that he wouldn't continue to throw barbs at would-be gurus and psychic charlatans but, like learning a new language when you hear it all around you, after finding himself a resident of California he set about becoming fluent in spiritual matters, as an insider rather than an observer.

For millennia, looking inward has offered an alternative to unpleasant circumstances if not horrors in the world outside. By early 1942, around the time the Huxleys were scouting for a getaway home in the high desert, Huxley's friend Gerald Heard invested an inheritance by acquiring three hundred rural acres located seventy-five miles south of Los Angeles. His plan was to build a spiritual retreat, a center for personal peace, for meditation and contemplation of the message at the heart of the world's religions, their mystical core. Aldous helped Gerald formulate a practical plan for the retreat and Heard called it Trabuco College. Together they crafted a mission statement (now preserved in the UCLA library Department of Special Collections).

During the war years, people some might call pilgrims traveled thousands of miles to Trabuco College hoping to meet its founder. Many had read Gerald Heard's 1939 book *Pain, Sex and Time: A New Outlook on Evolution and the Future of Man*, in which he expanded on his theory of the evolution of the human mind. Some arrivals sought dialogue with him, others yearned for a spiritual retreat, and many of them sought both outcomes. Heard gave lectures in which he blended a smooth synthesis of the world's sacred scriptures with his own incisive interpretations, sprinkled with quotes from spiritual leaders, but that recipe did not suit everyone who came to Trabuco.

The original buildings remain intact today— a brick bell tower, a red stone cloister with a covered walkway—all dating from the early 1940s when it was called Trabuco College, though the monastery is now operated by the Ramakrishna order of the Vedanta Society of America. On the day I saw Trabuco, the paneled monastery library with its big beams, fireplace, and the fragrance of wood smoke seemed unchanged from the days when Heard, Huxley, Christopher Isherwood, and many others gathered there six decades ago. Aldous stayed there on many occasions, and at one point in 1943 he remained for several weeks at what Maria then called "Gerald's monastery." The peaceful yet austere brick-and-stone retreat in many ways embodied Huxley and Heard's prewar pacifist vision, and I suspect that it evoked memories of Huxley's CO refuge during the Great War at Lady Ottoline Morrell's manor house, Garsington.

During my visit to Trabuco in the autumn of 2009, one of the first things I did was take a walk along what is called the Shrine Trail. Every few steps I came upon a marker with a symbol of one of the major religions. Around me oak leaves rustled and the fragrance of sage floated in the air. The path meandered along a steep drop-off where, if I squinted from where I stood at the top of the canyon, I could see the blue smudge of ocean edging Laguna Beach about seventeen miles away.

A crunch of footsteps told me someone was approaching on the trail, so I turned to see a young robed monk in his thirties with what struck me as incongruously reddish hair. He greeted me, the only visitor, and we started a conversation. I told him of my interest in Aldous Huxley and his friend Gerald Heard, who had been Trabuco's founder.

"Our swami met Gerald Heard in the early days," he said. "Swami is here, if you would like to meet him." The young monk led me through a shaded cloister walk past two large sleeping dogs until we reached a heavy wooden door. Out walked a thin, eighty-something Caucasian man wearing a robe, a wooly cap framing his kindly face.

Swami told me he had not personally met Aldous Huxley, but he had read and admired both Huxley and Heard's books as a young man. After fulfilling his alternative service on the East Coast as a conscientious objector, he told me, he had made it a point to meet Gerald Heard. Like many others, he traveled across the country to Trabuco College, where he found two dozen people engaged at that time in the study and practice of the mystics. Later I would learn from his biography that some of them also were involved in high-level experiments with ESP.

As I listened to Swami, I sensed a slight hedge in his explanation and, when I tried to probe a bit, he suggested that I read a published interview with him that explained more about that early period at Trabuco. The red-haired, robed assistant gave me a card with his e-mail address. A monk with e-mail, I thought, something unimaginable in the founder's time.

After my visit to Trabuco, I tracked down Swami's interview and found a copy of his book *Six Lighted Windows*, a spiritual biography of six Vedanta swamis with a thread connecting their teachings to his own. I read in his brief biography how he had fulfilled his conscientious objector's wartime duty by working in a psychiatric facility before coming to California. Then, too, I got the impression that many who made the pilgrimage to Gerald's monastery found him more the lecturer and former BBC announcer and less the spiritual guide. The swami I met at Trabuco said he gave up on Heard's curriculum and took up the discipline of Vedanta. Years later he would be known by his spiritual name, Swami Yogeshananda, and return to preside over the monastery where for him it all began.

During the later war years of the 1940s, around the same time as the young man who later became Swami Yogeshananda set out on his journey to Trabuco, another young man who would meet Gerald Heard, and through Gerald, meet Aldous, also made the same journey. Unlike the Swami, this individual would become a member of Huxley's experimental circle of friends. Both young men had read, and were impressed by, Heard's *Pain, Sex and Time.*

Huston Smith, with a recently minted doctorate in religion, was soon to make a change from his last academic post in Denver to his next position in the Midwest. In the interim he made a side trip to California and found his way to the top of Trabuco Canyon, where he spent one day and one night, much of it in intense conversation with the founder, and at the end of their time together, as Smith recalled in his memoir, Gerald asked him, "Have you met Aldous Huxley?"[3]

Huston had hardly imagined encountering this legend face to face, but in about the time it took to make a phone call Gerald had arranged a meeting. It only remained for Smith to find his way, not an easy task, to the Mojave Desert. A day or so later his bus pulled up in front of a frame house near a wide spot off the highway. As soon as his feet met the dusty ground of Highway 138, Huston looked up and saw Aldous on the porch waiting for him. He says he could barely believe his luck: Here he was, the invited houseguest of Aldous and Maria Huxley, who turned out to be "as friendly as can be," a sophisticated, urbane couple living in a setting both stark and beautiful in a seeming desert-of-the-Old Testament. No doubt this historical allusion occurred to the professor of religion who had been raised in China by missionaries and was a specialist in Judeo-Christian thought, although Huston Smith's spiritual orientation was soon to change.

During the time they spent together here in a place called Llano, the older and younger man, both seekers in their own way, took walks in the pink light of a desert sunrise and in the amber light of afternoon. Huxley, always a font of quotations, may have recalled the words of Goethe: ". . . how the gravity of Nature and her silence startle you, when you stand face to face with her, undistracted." Their conversation turned to the teachings of the Desert Fathers, hermit monks who lived in Egypt around the third century, underwent harsh ascetic practices in preparation for what they anticipated would be an encounter with God.

On their walks around the Huxleys' ranch, Huston and Aldous sidestepped creosote bushes and desert holly, and their conversation touched on Gerald Heard's belief that humankind had the potential for an evolution of consciousness. They extended that topic to the practices that may bring about such an evolution, variously called meditation, contemplation, and prayer.

Such matters would have been on Huxley's mind at the time. The subject of his 1941 nonfiction book, *Grey Eminence*, was well-born, seventeenth-century Capuchin friar François Leclerc du Tremblay, who became known as Father Joseph. Early on a practicing mystic, and said to be a healer, Father Joseph became infamous for his ruthless exercise of power after being recruited as the right-hand man of Cardinal Richelieu, the chief minister of King Louis XIII. The epithet "éminence grise," referring to the color of Father Joseph's robe, means the one in command behind the person seemingly in power, the man who actually pulls the strings.

Huxley had recently begun a new book about the mystical

thread in all religions, and that day in Llano he was primed for a wide-ranging conversation with a visitor who was about to take up his next teaching post at Washington University in St. Louis. Before his guest left the ranch, Huxley recommended that Huston arrange to meet the swami who presided over the Vedanta temple in St. Louis, though at the time the young professor admitted that he didn't even know what the word swami meant.[4]

Huston Smith would recover brilliantly from this omission and go on to become the foremost authority on comparative religion. His book *The Religions of Man* would sell more than 2.5 million copies in different editions after its publication in 1958. That same year he would serve as liaison for Huxley's semester as a visiting professor at MIT, leading to the first meeting between Aldous and a newly arrived young Harvard professor by the name of Timothy Leary. But wait, I'm getting ahead of my timeline.

After Pearl Harbor, after Roosevelt had declared war on Japan and the Axis powers, the Huxleys had traded the coastal climate of West Los Angeles and the complexities of Hollywood for a simple and serene desert habitat. They brought with them an appreciation of nature and a playful love of picnics. Initially they may have heard about this place called Llano from the film community and decided to take a look. Or curiosity might have stemmed from Huxley's reading, or likely from Maria reading aloud to him about potential desert locales for their next temporary home, and this one came with both legacy and affinity.

One day while driving on a sun-bleached two-lane road connecting several barely populated Mojave towns, they came upon an old turn-of-the-century frame house with a For Sale sign. I imagine Maria's foot pressing down on the brake and a puff of dust curling around the tires as they considered the shabby state of the house. Aldous, a font of historical lore, this man who took Firsts in English and History, might have looked over at Maria and said something about the utopian community adjacent to the property. Once it had stood proudly but had left only scattered ruins behind. This was once the Llano Del Rio Cooperative Community.

With his ready memory for details, he might have gone on to explain how Llano had been founded forty years earlier by a former socialist candidate for governor of California, that it was a commune built of local granite boulders with grandiose plans. Its receptive soil enabled the self-sustaining community to manage until various problems, including an internal power struggle and artificial limits on water, eventually brought it down.

Now only a few stone walls, a raggedy tower, and a pair of rugged fireplaces remained. Aldous might have observed that the founders aspired to an earthly paradise, but that any utopia harbors the seeds of a dystopia. Or maybe he reflected on his involvement in a more recent failure of intent, the ill-fated Peace Pledge Union. In any case, for the sake of a dry climate to benefit his own and Maria's health, and as a place far away from city hubbub where Aldous could write his next book—or as it would turn out, several books—they decided to purchase this getaway ranch that would turn out to be both trash and treasure.

After the sale closed it became apparent that the place was more deteriorated than supposed in their first flush of enthusiasm, but it became more comfortable after paint and repairs, and they added a two-room studio for Aldous. The arid landscape outside his window provided an unlikely source of inspiration, one geologic configuration being a hummock that locals call the Sleeping Elephant.

Aldous had taken up the Bates method of eye exercises four years before moving to the desert,[5] and now Llano's glassy light aided his vision and made him keen on devoting more time to this discipline. He delayed work on another novel, devoted himself to his eye exercises, and set about completing a book about eyesight that evolved into more than an essay about what the eye perceives. You might say it was about both insight and outsight, foreshadowing a pair of controversial books he would write a dozen years later. Comprised of external observations interwoven with inner reflection, *The Art of Seeing* would be published later that same year.

Other projects during the desert years included his novel *Ape and Essence*, which opens with a description resembling Llano. He would finish another novel, a book of nonfiction, and write a failed screenplay in collaboration with a fellow expat, Christopher Isherwood, who would later be known for, among other published work, the *Berlin Stories* that became the basis of the 1960s stage and film hit *Cabaret*.

Llano would remain the Huxleys' home base until 1945, when they reluctantly moved to a small home in the nearby mountain community of Wrightwood, but the Llano ranch with its utopian ruins had more literary resonance. The white frame structure was nestled in an oasis of mesquite trees near Big Rock Creek, and although it was barely more than an irrigation channel the *rio*'s effect was, intermittently and seasonally, like a ground-seeping spring. Grape vines wound around the house, trees provided shade and fruit to eat, including plump figs. Even on their first visit to Llano the Huxleys had noticed how the irrigation ditch served as a watering hole for thirsty wildlife. Ensconced in their desert hideaway and far from constant reminders of war, they adapted to their own fragile utopia.

I picture Aldous in this setting, walking across a plank bridge over the irrigation channel, his tall, lanky frame bent slightly as he ties supports onto a bean plant. Here among the ripening tomatoes and assorted vegetables Aldous continues the discipline of the Bates method, which involves focusing close-up, then more distantly: close-up for tendrils of green beans, more distantly for the branching parent stalks. As his eyes continue to improve in the clear desert light, he learns to drive and even takes the wheel alone, limiting excursions to unpopulated roads where miscalculation simply means a collision with sagebrush. Unobstructed natural light streams into a study purpose-built as his workplace, supporting his working vision, at its best since before a disease ruined his eyesight. Far removed from tall buildings and smog, he finds he can read for short periods without wearing glasses.

They thrive here, up to a certain point in time—the garden, early morning walks when evaporating dew carries the scent of herbs, sometimes a walk in the afternoon before the sun dips below the horizon and golden light turns into long blue shadows. For Maria's part, her benefits, her gifts, during the years when Llano was their primary home are twofold. Her lung condition improves, but the second gift is of an unexpected and higher order. Maria's boundaries seem to dissolve in this place. She comes to call this transcendent state "my magic," and a few years later, after hearing Aldous describe his first mescalin experience, Maria will say, "I have known that it is like this all along." In the desert, Maria knew the visionary experience spontaneously.[6]

At night the howling of coyotes could wreck a good night's sleep, and the eye exercises turn out to be of limited benefit in the end, but Aldous is a person not quick to give up on healing. Here in Llano, healing is revealed as a much more mysterious process than he had ever imagined.

Wartime rationing of auto tires and gasoline limited back-and-forth travel, so Aldous sometimes stayed for a stretch of time at Gerald's Trabuco College. Such a stretch came in July of 1943, while he worked on two manuscripts. One was *Time Must Have a Stop*, with its theme of limbo and regret and the lost chance of finding a meaningful life before death closes the door on preparation. This book, which Aldous will later say was his personal favorite among his many novels, contains the idea of the Minimum Working Hypothesis: Without a hypothesis, there is no reason to go forward in research. Religion is research. Assuming the existence of a Ground of Being or a Godhead or a Clear Light, or whatever you want to call it, serves to motivate experimentation in spiritual experience. After first writing about this in an essay in a 1945 issue of the publication *Vedanta for the Western World,* Huxley incorporated it into a chapter of his novel *Time Must Have a Stop.*

Another work written in part at Trabuco and in part at Llano is *The Perennial Philosophy.* This explanatory book, woven with quotations and blended with commentary, is another of Huxley's collections to be dipped into for sipping, plumbed from the underground stream coursing through major religions where Huxley taps into the thermocline layer of mystical tradition. Or as William James once made the distinction: to most people, religion offers a way of life; to the seeker, religion offers a way *to* life. As a travel guide, not Huxley's first nor his last, *The Perennial Philosophy* has a wayfarer's feel about it.

The July 1943 period at Trabuco when Aldous worked so productively brings to mind how beastly the Mojave Desert must have been in summertime, though Britons I've known seem to relish such extreme weather. Trabuco College was at a higher elevation and more arboreal. Huxley described Trabuco in a February 7, 1947, letter to Isherwood as "a huge estate in a very beautiful rather English countryside behind Laguna." But it could turn miserably warm. When the students and pilgrims and friends residing at Gerald's monastery longed for ocean breezes, that desire sometimes called for piling into a car and making a visit to the blue smudge I saw from the Shrine Trail, the seaside town of Laguna Beach.

The passed-down lore of this beach enclave, founded as an art colony in the 1920s, documents one such a wartime visit as recorded by city historians. As the story goes, one day in the 1940s a civil-defense volunteer canvassing the town knocked on the door of a Laguna Beach residence. The volunteer must have been someone with a literary bent or this story would not have survived because it seems that when the door opened he, or maybe she, saw two or three notable British authors: Aldous Huxley, and with him Christopher Isherwood and probably Gerald Heard, too.[7] I picture Huxley's long legs stretched out for comfort, Aldous being tallest of the three. The volunteer stood in the doorway as the identity of these authors, perhaps recognizable from photos on book jackets, dawned on her and she probably wondered why they were all there together. They may have been taking a break from summer's heat pooling inside the red brick walls of Gerald's Monastery.

As the war effort phased into later stages, my father's work changed. After designing and teaching his classes, and after his textbook was implemented in Engineering, Science, and Management War Training classrooms, he was reassigned to supervise draftsmen working in a huge space in the Vultee aircraft plant, where fluorescent paint on floors and walls let employees work under blackout conditions twenty-four hours a day.

Many of the men suffered crushing headaches after repeated exposure. My father's own headaches may or may not have been related to the glowing paint, but doctors could come up with no diagnosis for an intractable pain so severe that he thought he might lose his job, if not his mind. Then one day my grandmother heard about a so-called spiritual healer, a man named Dowling. My mother contacted him and made an appointment, and my father reportedly grumbled, "What the hell," thinking he had nothing to lose. A day or so later my mother drove him to the appointment.

I've been told this is what transpired next: that my father reached the front porch and when Dowling saw him the man clutched at his own head with an apparent cry of transferred pain. He motioned for my father to come close, then began a continuous flowing movement of his hands, moving them slightly and rhythmically, hovering near his subject but not touching him. In a few minutes the crushing headache was gone, and it didn't come back. To my father it seemed just short of miraculous, as if the healer had somehow isolated and dispersed—banished or exorcised—his pain.

The next day my father returned to Dowling and said he wanted to learn his technique and offered to pay. Dowling said he accepted neither students nor payment, but my dad kept showing up after work. Persuaded of my father's serious intent, the man relented and allowed him to watch but would not permit questions, as if the silent treatment would drive the petitioner away or else allow all to become clear. This was how my father's after-work apprenticeship as a healer began. "How you open the door is the trick," my father once told me. "The rhythm of your hands drives out the words."

In 1944, while living at Llano, Aldous began writing screenplays with fellow expat Christopher Isherwood. Aldous suggested that the two writers collaborate on a screenplay based on an individual who allegedly lived near the Huxleys' ranch in Llano, an old man, the locals said, whose hands had the power to heal.

What appealed to the two writers was the idea of a man living a modest life, who had somehow evolved in consciousness to the point that he had harnessed an extraordinary and improbable energy, a power outside ordinary standards of measurement. By this time Isherwood had become involved with the Vedanta Temple in Hollywood and both he and Aldous were spending time at Gerald's retreat in Trabuco. Huxley had also become friends with Krishnamurti, and several of their mutual friends practiced meditation. The time seemed right to plunge into such an unconventional project, a drama with an unorthodox spiritual theme, a screenplay they came to call *Jacob's Hands*.

But after Aldous and Christopher had completed a draft they deemed worthy of submission, the studio decision-makers balked. Huxley's name was a potential draw, but he was best known for his wry, satirical novels and this screenplay smacked of, well, sincerity.

Add the rule of thumb that a wartime audience wants heroism, romance, and escape, and this healer character had to be young, handsome, and possessed of a fire-and-brimstone personality, plus the story required a compelling love interest. They did a treatment with Jacob as a young man and crafted a romance, but in the end the script was rejected. Huxley wrote Isherwood after hearing the news from the William Morris agency. "It appears," Aldous wrote, "that the reason for the hitherto universal rejection of it is fear of the doctors."[8] There might have been the additional problem of an implied paganism, of showing spiritual healing disconnected from Christian faith. For any number of reasons the screenplay ended up in Huxley and Isherwood's respective bottom drawers. One copy would be destroyed by a fire. Its twin, forgotten in storage, would not see the light of publication for another fifty years.

In an equally if not more frustrating outcome, the time at Llano did not have a good ending. The Huxleys' idyllic life on the ranch came to a halt after a profusion of ragweed, exploding in some recurrent cycle, caused an allergic reaction that sent Aldous to the hospital. This was sadly ironic since the prospect of breathing pure air was what had brought them to the desert in the first place. The diabolical plant so stubbornly resisted eradication that when faced with this enemy the Huxleys could only escape. They left their Llano house empty for a while and moved to a higher elevation in Wrightwood, where they found a small place surrounded by pines and arranged for a silver trailer to be parked outside to serve as the author's studio. They even managed to lure a few of their friends, including Krishnamurti, to visit them in Wrightwood and stay under the pines for a while.

For around a year they hung onto their empty desert home sur-rounded by fruit trees and a now-neglected garden, but they would never live in Llano again. In a letter around the time they had to abandon it, Maria described the deep sense of loss they both felt over leaving their desert utopia turned dystopia. Maria wrote of her "pangs of regret for the beauty" and asked, "Should it be allowed to play such a part in one's life?"[9]

The war had hijacked almost everyone's plans, and in many cases so did the war's end. Gerald Heard gave up Trabuco College five years after founding it, deeding the debt-ridden property to the Vedanta Society of Southern California. By the time of the 1947 handover, maybe Huxley and others felt less a need for this spiritual retreat, which had been founded on the eve of a global conflict.

Still, Huxley and Heard made occasional visits to Orange County's seaside town of Laguna Beach even after Gerald relinquished Trabuco's keys to its new owners. If you were driving from LA to Laguna in those days, heading south on Coast Highway, you passed oil wells pumping like the bobbing mandibles of giant insects. You passed a scenic auto pull-out with a semicircle of cars parked around a hut selling date shakes. A mile or so farther you entered downtown Laguna at an intersection presided over by a shaggy-haired Dane named Eiler Larsen, who waved a greeting and called out a cheery welcome to all who entered.

Every Sunday in that postwar period our family of three routinely drove the twenty-eight-mile route from Long Beach, on the southerly end of Los Angeles County, to Laguna Beach in Orange County. When we reached the intersection I waved back to the Greeter (identified in my dad's collection as Larsen of Laguna, hand photograph number 239). This was not just a leisurely Sunday drive. With industrial work hours sharply reduced following the war and more available time for experimenting with healing and his hand project, Dad was on a quest to find answers. Did the visible patterns in the hand have anything to do with *healing* hands? Was there a wellspring, an invisible source of power at play, like the electrical current you plug into different appliances? The way he approached these questions was to seek initiation into an obscure religious order. Its followers met in a small brick church, built out of earthquake rubble, consecrated in the name of St. Francis. Later it would come to be called St. Francis-by-the-Sea.

It was late fall in 2001, and Dad and I had been meeting in the patio behind my grandparents' bungalow for about two months. In some ways it was like a time warp. My father had left our family in 1957, remarried a few years later, and as a widower circled back to the scene of his first marriage. He'd even tried to flirt with my mother, but she kept diplomatically ignoring him.

The other unexpected thing was the way the boxes dredged up memories for both of us. Dad and I had already sifted through four batches of hand-related photographs, so on that day I brought out the box with its handwritten label, The Church. I opened the lid and my eighty-two-year-old dad, quite irreverently, considering the words on the label, said, "Well, I'll be damned."

His blue eyes sparked with interest, and since I'd already looked inside I knew that the contents dated from Dad's participation in the arcane order whose members practiced their rites inside a scaled-down cathedral so small it fit on half a house lot. Tiny St. Francis-by-the-Sea had even been acknowledged in the *Guinness Book of World Records* as the "smallest cathedral in the world," defined as a church designated as the seat of a bishop. And since this was the only church of its denomination and it had its own bishop, it claimed the honor. Now listed in the National Register of Historic Places and mainly used for weddings, back when my father became one of its so-called adepts the church had existed for slightly more than a dozen years.

From the corner of my eye, I saw him watching me as I removed the first item from the box. It snaked out like a magic trick, a narrow purple scarf of heavy satin, with a tendril pattern embroidered in gold.

"What's this?" I asked. It did look familiar, and maybe it was something I'd seen on Sundays back when Dad was part of the procession and one of the ceremonially robed men.

"This was one of the vestments," he said.

Next I reached into a bag made of burgundy flannel and pulled out a brass chalice. I remarked on how well the cup had maintained its shine despite being untouched by metal polish for nearly fifty years.

"The bag is treated with the same type of preservative used for storing silverware," he explained, always the practical science guy except when he touched on topics many would call pseudoscience. He'd spent much of his adult life trying to span these two polarities, and he believed it was only a matter of time before technology made the improbable probable, translated the uncanny from anecdote to analysis. After his own fashion, this was what he'd been trying to do for many years.

The unpacked trove with its glitter of brass and sheen of satin was starting to fill the patio table. The next object I removed from the box was a roll of nested parchment scrolls. The outer scroll, dated 1948, bore the signature of someone with the title Grand Master. Reading the inscription, I learned that my father, Howard Alban Thrasher, was an initiate (which accounted for this unfamiliar middle name) in the ancient order of Saint Louis, and the scroll listed one of his official tasks as the role of exorcist.

"Jeez, Dad, did you ever officiate at an exorcism?"

"Well," he said, "Yes and no. It had to do with healing."

While my father and I were discussing the things we'd rediscovered in the box, my usually cheerful mother emerged from the bungalow carrying a pitcher of iced tea. Her dark brown hair was newly coiffed from a weekly salon appointment. I had inherited my father's blond hair along with a pale copy of my mother's smile.

"Would you like a refill?" she asked, looking around for an empty spot to park the pitcher, then she recognized the objects on the table and a cloud crossed her face as if a sad memory of the lost years of their marriage.

We accepted her offer, and after she returned to the house I unfolded a roll of parchment showing a chart tracing the lineage of this curious church. The scroll read that the basis of the priesthood was something called the Old Law. My eyes landed on a paragraph, and I read it aloud:

> All ancient religions were essentially systems of natural magic, theosophical, theological, philosophical, and astrological. These systems of magic were inculcated by "Initiations" and classified broadly as the "Greater and Lesser Mysteries."

Tucked inside the nest of scrolls was a smaller sheet of paper, with the inked outline of a hand drawn in my father's precise draftsman's style. In this rendering, an astrological sign appeared on each fingertip like a tattoo. It made me think of the passage I had just read, about mysteries one portion astrological.

I felt a passing breeze like a memory of my occult childhood, a time when my parents held séances in our living room.

Re-nesting the parchment and trying not to sound skeptical, I asked, "Are you at liberty to tell me about these mysteries, or are you still sworn to secrecy?"

"I am free to tell you," he said, "but it is up to you to understand."

I suppose the parishioners who attended those Sunday services at St. Francis-by-the-Sea didn't stumble in by accident. They knew its unorthodox rituals had to do with ancient mysteries, and that its bishop had the keys to some kind of occult toolbox. But to a seven-year-old kid it was just an hour-plus of pure boredom. I remember one day in particular. My mother and I were in our usual place on one of the smooth wooden pews, while jewel-like colors streamed through the stained glass windows.

When the bishop appeared, clad in a white robe with a purple cape and a peaked gold-and-white cap, looking portly and regal, a hush fell over the fifty or so people seated in the small chapel. From a vestry in the rear, into which my father had disappeared a half an hour earlier, a blond head emerged and now Dad wore a white robe with a purple satin stole, embroidered with gold vines, draped over his shoulders.

He walked up the center aisle toward the altar, its side panels framed by filigreed screens. When he passed us, a stream of smoke billowed from his censer as he swung it left then right, double right then double left, leaving a wake of incense behind.

As I come out of my weird reverie, Dad asks if I remember the old church and I say of course, and it occurs to me that more than half a century later I am writing about the World's Smallest Cathedral, and soon I will be writing about the World's Biggest Drugstore.

Around the time Dad and I delved into the box labeled The Church, I heard about a posthumously published work by Aldous Huxley and Christopher Isherwood. The initial report had appeared a few years earlier in the *Los Angeles Times* (on November 9, 1997), after actor Sharon Stone saw a mention of a rejected script in Christopher Isherwood's *Diaries*. At the time, she had contacted Laura Huxley and expressed an interest in seeing the story—called "Jacob's Hands"—and, although her interest did not result in a film option, it led to the story's rediscovery.

Actually, it was an unproduced scenario with dialogue and ideas for scenes. Huxley's second wife, Laura, along with Isherwood's long-time partner, the portrait artist Don Bachardy, had collaborated to have it published as a novella called *Jacob's Hands: A Fable*. This is how I learned that the work written when Aldous lived in the desert had finally ended up in a published form. When it arrived in bookstores in the late 1990s I had not yet found the boxes so it was not on my radar screen, but hearing about it in 2001 I was eager to find a copy. I located one and met with Dad on the patio.

I set the paperback on the table. "Do you know anything about *Jacob's Hands*?" I asked him. "It was written by Aldous Huxley and Christopher Isherwood, and it's about a ranch hand with the power to heal."

He raised an eyebrow. "Well, Isherwood once invited me to his house in Laguna. We talked about my healing work."

It seemed incredible at first but in a way that made sense, for otherwise my dad was the last person likely to be invited to the home of a literary figure like Isherwood—except that by now I knew my dad had spent time with Isherwood's friend Huxley.

"Do you remember where Isherwood lived?" I asked, trying not to sound too skeptical.

He thought for a moment. "The house was on a hillside in South Laguna," and I thought, but didn't say, *many houses are on hillsides in Laguna.* I could tell he was water-witching into long-term memory. "It was on a street closely parallel to Coast Highway." That narrowed it down.

"So, Dad, do you recall when this took place?" I probed, trying not to let disbelief creep into my voice.

"I was involved with the church at that time . . . so I'd say about 1949 or 1950. Do you remember my ink drawing of the hand, the one you found in the box?"

"The one with astrological symbols like tattoos on the fingertips?"

"If you turn the drawing over you will see the words: 'I exorcize all influence of evil.' I read it during the anointing ceremony. The purpose was healing, and that is the kind of thing we talked about." I asked if he remembered anything else about Isherwood's house, and without hesitation he said, "It was up a steep walkway on a street off Coast Highway. It had a terrific ocean view."

I felt compelled to check this out, wanting proof that my father was a reliable interviewee, so I tracked down Isherwood's published diaries for that period,[10] checking in the index for places he had once lived. One house he rented in Laguna Beach fit the time period. The index mentioned an address, so I drove to Laguna and cruised the street, squinting at mailboxes for house numbers, checking up on my dad's veracity or perhaps his memory, and it turned out he knew what he was talking about. I found the number, and the house was era-appropriate for 1950, located on a street parallel to Coast Highway. It stood on a hillside with a steep walkway leading to front windows that still framed my Dad's long-remembered ocean view.

The Huxleys' son, Matthew, said his parents' early existence in America resembled the life of gypsies, according to biographer Sybille Bedford. As it turned out, the ragweed that forced Aldous and Maria Huxley to flee Llano for the mountain respite of Wrightwood would take them a step closer to owning a permanent residence. But what came first was a return visit home. Luxury ships converted to carry military personnel and supplies were back in passenger service, and the Huxleys decided it was time for their first visit to Europe since arriving in New York under the impression that it would be a short stay.

Eleven years had passed since their arrival, and their return would be made on the RMS *Queen Mary*. Launched just before the war, and soon converted to a troop carrier, the ship's blonde wood, murals, and etched crystal had been restored to Art Deco luxury. After a smooth crossing they visited family members in Britain and Belgium, and while in France they began the process of selling the house that had been their last residence abroad, the place with a typo on its sign, Villa Huley, in the seaside town of Sanary-sur-Mer.

Aldous had another travel destination in mind for research. The French village of Loudun was the locale of an infamous case involving diabolical possession and rites of exorcism. The well-documented seventeenth century incident, rich with details both carnal and theological, would be the subject of Huxley's next nonfiction book. He would delve further into the topic of transcendence gone awry that he had explored in *Grey Eminence,* his biographical study of Cardinal Richelieu's mystic turned foreign secretary, except this new project would veer in the direction of mass hypnosis and madness.

After returning from their trip abroad the Huxleys began looking for a more permanent home. They found a property in the Spanish Revival style of the 1920s located just off Melrose Boulevard. Its address was 740 North Kings Road, Los Angeles 46, in an era of double-digit zoning before the innovation of zip codes.

The facade and inner walls, and the courtyard with an arbor, reflected a boom era when faux architecture had flourished in Southern California. The Moorish architectural details had a warm association for the Huxleys because during their early married years they had spent extended time in Spain. This house with its foot-thick adobe walls and lush foliage recalled the walled gardens of Andalusia.

Apart from a muted natural light and residual dampness from those same lush gardens, the house enjoyed the advantage of location. For the Huxleys' purposes, the house was near the well-known Farmers Market, not far from the film studios, and within blocks of a sprawling Owl Rexall that soon would become a symbol to Aldous of the diverse oddness of America's consumer culture. Displays in the drugstore's football-field-size interior offered every imaginable remedy, gadget, implement, trend, trinket, gee-gaw, and curious doodad, all under one roof. It was known by its slogan as the World's Biggest Drugstore.

Soon the Huxleys had established a new routine, with regular visits to libraries and bookstores, including jaunts to see rare book dealer Jake Zeitlin, who was helping Aldous track down books for his possessed-Ursuline-nun project. The Huxleys made trips to the sprawling Farmers Market with its open-air fruit and vegetable stands and the adjacent Town & Country shops and cafes, where Aldous and Maria often had lunch with friends from the disparate realms of science, music, and film.

Around that same time, Aldous began hosting weekly gatherings for a very different group of friends.

On Tuesday nights when the Huxleys were in town, Aldous invited guests whose names and interests were often made known to him through other British expats. This was an unusual and non-literary salon. You could say it was based on the one thing the guests had in common: the exploration of phenomena that current science could not explain. Especially of interest to Aldous was the unexplored capacity of the human mind.

My father met Aldous around 1951 through the British ex-pat network of Isherwood, Heard, and Huxley, with their shared interest in spiritual and parapsychological pursuits. One of Dad's friends, a fellow adept from St. Francis-by-the-Sea, lived in Santa Monica and, as it turned out, his home was a few doors away from Gerald Heard's.

"I had no idea this was going to lead to Huxley," Dad said, recalling the day he was visiting his colleague, who happened to mention his famous neighbor. My father's friend also mentioned that Heard was involved in the Society for Psychical Research, and since my dad happened to have a few hand photographs with him that day he decided to walk over and ring the bell.

"A housekeeper came to the door and told me 'Mr. Heard is not at home,' but I had a feeling he was," my father recalled, "so I gave her one of my photographs and said, 'Would you please give this to Mr. Heard.' I waited on the front porch for a minute or two. Pretty soon, just as I'd guessed, he poked his head out and said 'Please come in.'"

As the two men talked, Dad explained what he had been doing

with his hand study, which was analyzing structural properties of the hand in relation to talent, personality, and other traits, among them mental illness.

"Huxley's wife collects hand prints of famous people," Heard said, peering closely, "but her prints are nothing like these." Those of Mrs. Huxley, my father learned, were simple ink impressions in which the fingers appeared like narrow stripes and the palm appeared sketchy, with large areas blank and unrecorded.

Heard said Huxley was also interested in typologies, or studies of possible correspondence between physical manifestations and human traits. Huxley's pet typology was Columbia professor William H. Sheldon's human somatotypes—the *mesomorph type* (a muscular/square body suggesting a physically active individual), the *endomorph* (fat/round, more socially inclined), and the *ectomorph* (thin and often tall, usually a reserved and cerebral individual).

"Gerald arranged to take me to Huxley," Dad said, "and once I was there I got in a discussion with Aldous, who was also interested in spiritualism, and I was into it up to my ears with the church in Laguna. He started probing, and since there was no reason for me to hold back I told him about several of my experiences. Aldous could see that I had created something different with my hand project, a system, a study with depth, so he said, 'Suppose we hold a class once a week at my home.'"

CHAPTER 4
CHASING SCHIZOPHRENIA

Schizophrenia affects one percent of the world's population. Despite the symptom-mitigating power of antipsychotic drugs, the affliction is far from understood. It may seem hard to believe, looking back from our smug perch in the pharmaceutical-heavy twenty-first century, but a few decades ago the notion of a biochemical basis for mental illness was a radical, and even a ridiculed, idea.

From the 1920s on through the 1960s and even into the 1970s, it was widely accepted that familial influences were the basis of mental illness. The idea persisted that bad parenting was the main reason for schizophrenia; maybe a distant father, but primarily a "schizophrenigenic mother" was assumed to be the cause.[1]

Today, Freudian psychoanalysis and the school of psychology called behaviorism are practiced in greatly modified forms. A Freudian psychotherapist heals by making the unconscious conscious, including the use of dreams. Behaviorism died and was reborn as part of cognitive behavioral therapy, now practiced by many leading therapists. But these were used far more narrowly in the early 1950s, when Freud ruled as the secular god of psychiatry while—following the work of J. B. Walsh and B. F. Skinner —behaviorism dominated the field of academic psychology. According to the tenets of behaviorism, our thoughts, hopes, and impulses are largely attributable to stimulus and response, reward and punishment, not unlike Pavlov's famous experiment with the salivating dogs.

In the 1950s, however, the gradual development of new drugs would undermine this view.[2] Among the first wave of psychiatrists to challenge the establishment by arguing for a biochemical basis for mental illness were two British doctors, Humphry Osmond and John Smythies. The two men were different in many ways. Smythies, five years younger, had been drafted into the British Navy in 1945 as a Temporary Acting Surgeon-Lieutenant in the post of ship's doctor as the war wound down. Osmond was reassigned after convoy duty to the base hospital on the Mediterranean island of Malta. He had not yet met Smythies, the future colleague who would be responsible for bringing he and Huxley together.

By temperament—and he admitted as much—Dr. Humphry Osmond liked a challenge, a good conceptual dust-up. Smythies described him as "a tough and stocky Scot with a keen intelligence and a remarkable range of interests,"[3] whereas Smythies was a tall physical type somewhat like Aldous.

Between the two eventual research partners, John Smythies had the most compelling personal reason to challenge the psychiatric establishment. As he wrote in his memoir, during his last week as a general surgeon trainee an uncanny experience altered his course. He had just undergone the end of a personal relationship and was alone one night, when he felt his life had reached a new low. That same evening, Smythies had reluctantly agreed to accompany a friend to a religious service, though actually this activity was not of much interest to him for, as he recalls, "I had just lived the ordinary pagan life led by most of my fellow students, who listened to Darwin, Marx, and Freud." He went along for lack of anything better to do, and after a service that left no particular impression he returned to his rooms and felt as if he was being ground between the millstones of regret and remorse. Then suddenly the millstones stopped. "I looked over to the corner of the room and saw a small oval of clear white light," he said. "I felt enveloped in a great peace and closed my eyes and went to sleep. I woke in a state of joy and realized that something momentous had happened."[4]

Smythies wrote in his memoir about that night: "I came to understand that I had been through the experience of enlightenment described by the Buddha, St. Paul, and many others." The next day he mapped out a new direction for his life: He would pursue a cross-discipline grounding in philosophy, medicine, and the science of the brain. Although he still considered himself anything but religious, that window into a strange new reality on a November evening had left him so stunned that even months later he was still "fizzing from my religious experience."

If this experience was illusory, then why was it so powerful? How had it brought about an entire change of orientation? The force of one incident compelled him to seek explanations, and one avenue was a study of hallucinations.

In 1950 he transferred to London's St. George's Hospital in a three-man psychiatric unit. Around that time, he treated a patient in the emergency room and learned that the man was a professional hypnotist and clairvoyant who had recently performed at a nearby theater. Clearly this was someone who commercially gave public demonstrations of mind reading, but in the hospital he would have no access to tricks. When Smythies asked if he would be willing to try an experiment, the clairvoyant agreed. A day later Smythies handed the man a pair of sealed envelopes. The test involved providing details about the two people who had written messages on sheets of paper inside, while also identifying their locations. The specific words written by these two people were unknown even to Smythies. Before the envelopes were unsealed, the patient provided accurate details about the people and locations that someone limited by the constraints of time, space, and a hospital room could not possibly have known.

Convinced that this phenomenon was worth exploring further,

John joined London's Society for Psychical Research (SPR). He had not yet met Gerald Heard, another member of SPR, who was now living abroad. Upon joining the London SPR, John Smythies became one of its youngest members.

He continued to pursue his overlapping interests involving ESP and hallucinations, tracking down anything written on the subject. In 1950, six weeks after arriving at St. George's Hospital, he came across a book in French, written in 1927 by Alexandre Rouhier, who described the type of vision experienced by a shaman after ritually ingesting the flesh of a Mexican cactus called peyote.

In the late 1800s, pharmacologist Louis Lewin published a study of the peyote cactus,[5] and subsequently Arthur Heffter isolated mescalin as its psychoactive property.[6] The American philosopher William James—later chair of psychology at Harvard, as well as founder of the American Society for Psychical Research—had reported on his own transformative episode after ingesting a briefly mind-altering substance, nitrous oxide, which led to the Harvard lectures that became his classic 1902 work, *The Varieties of Religious Experience.* In the late 1920s, Heinrich Kluver of the University of Chicago investigated mescalin and hallucinations. This period was followed by a lull in interest during the otherwise preoccupied Great Depression and World War II years.

Smythies saw other implications, apart from accounts of natural substances giving rise to hallucinations. When he read about peyote visions in Rouhier's book he noticed that the author had cited the chemical formula of mescalin, the active principle of the peyote cactus. Seeing the formula in print, Smythies realized that this recipe resembled a biochemical compound found naturally in the human body: adrenaline.[7]

Smythies speculated that since taking mescalin induced a distorted visionary experience similar to schizophrenia, therefore a naturally occurring chemical in the human body might be related to schizophrenia. He conveyed this notion to a senior member of their three-man psychiatric unit, Dr. Humphry Osmond—who called it a fruitful idea worth pursuing. Together they set about searching for what they initially called "M-substance," alluding to its similarity to mescalin. But instead of encouraging them in their investigation, the ranking doctors in the psychiatric unit of St. George's Hospital showed little interest in this potentially new approach to schizophrenia, having no interest in research, much less any interest in metaphysics. As Smythies recalled, "In fact, we got the feeling that it was rather bad form."

My father's hand project had started in 1950, around this same time. Now it was half a century later, and lately I'd been sorting through Dad's box of numbered index cards, each one cross-referenced to a slide preserved in a wax-paper sleeve, wondering just how well the slides had survived after so many years of storage in the garage. They looked intact but I wouldn't know for sure until I'd had one printed at my regular photo lab, so I took the slide of Huxley's hands to be processed as a print and they turned out to be perfectly clear.

On our next patio day together I said, "So, Dad, by the time you'd started attending Huxley's Tuesday night meetings, it looks like you'd collected only about 250 photographs. I see that Huxley's hand is number 249."

"But by then I'd worked out the technique," he said with a grin. "It took me that long to perfect my camera and lights."

From the start he had been determined to measure every marker he could pin down, from finger length and nail dimensions to palm width, length, and thickness, along with capturing a photographic record of each hand. When he realized that flash bulbs washed out the fine features, he noodled the problem until he found a solution: print hands as negatives to make the patterns stand out.

"What was that minty stuff you rubbed on our hands?"

"I made an emulsion from tooth powder so the negatives were white against black," he said. "The contrast caught details of the lines and the grain of the skin."

I asked why he had bothered to expand the project to almost one thousand hands, and he told me that back in 1950 when he'd started the study he was convinced that a large sample had a good chance of showing some correlation between the hand and human traits.

"I'd sit by the hour and study the damned things," he said. "I told myself, *yes, there is something in the hands,* but I knew I'd have to have enough photographs, and different kinds."

Talking about the different categories that day on the patio became a memory duet—because I could remember some of them, too. He talked about taking hand photographs of different affiliate groups, and I remembered the times when he had brought me along. He collected photographs at a convention of identical twins in Huntington Beach, where I was told not to stare rudely at the duplicate faces, and where my mother wrote down measurements as he called them out. We photographed the hands of musicians in the Los Angeles Philharmonic Orchestra, whose members I guess I had expected to wear tuxedoes but it was a rehearsal day and the men were dressed like my dad in white short-sleeved shirts.

On other occasions I stayed home with my grandmother when my father and mother took measurements and photographs of the hands of machinists and people afflicted with stuttering. The Long Beach Convention Center drew a variety of professional and business associations and Dad, able to add convention groups with relative ease, captured the well-scrubbed hands of dentists, the manicured hands of the nation's top-selling insurance salesmen, and photographed what you could call more exotic specimens—the many-ringed hands of a convention of trance mediums.

Then he set his sights on a category of individuals that today a researcher, much less an amateur, would never have a prayer of obtaining permission to examine. He sent a letter to California's mental health officials in Sacramento. Remarkably, he was granted permission to study the hands of patients suffering from schizophrenia at two state hospitals, one group in Norwalk another group in Pomona.

Meanwhile, psychiatrists John Smythies and Humphry Osmond continued their search for the elusive M-substance. As Osmond would later describe it, they were looking for a biochemical found in the body that "might be involved in the production of the experiences schizophrenics endure in their distorted perceptual worlds."[8]

After finding their theory of no interest whatsoever to their supervisors in London, Osmond kept his eye out for a more congenial psychiatric hospital. Before long he noticed an item in the professional journal *Lancet* about a position in Saskatchewan, a prairie province in Canada. In this post, he would be working with a large number of schizophrenic patients and might be able to carry on more freely with his experiments. Moreover, the premier of the province, Thomas C. Douglas, was a social democrat keen on health care reform and an advocate of radical experimentation. Osmond applied and was chosen from among a field of sixty as the new deputy director of Saskatchewan Hospital in Weyburn, where Smythies would soon join him.

By the time Smythies was ready to leave London for North America he had begun to explore other fields of study. The strange phenomena he had experienced, his vision of the uncanny oval light, and the persuasive demonstration by the clairvoyant patient, had raised many questions in his mind about the nature of consciousness.

Phenomena such as hallucinations do not fit into the typical framework of how we perceive time and space. Smythies found a copy of J. W. Dunne's book *An Experiment with Time*, in which Dunne speculated that space has more than three dimensions. Over the next five years, Smythies would develop his own theory and publish an article called "Extension of Mind" in the *Journal of the Society for Psychical Research*, taking the stance that the mind is not equivalent to the brain and therefore is not physically limited to that organ's physical grounding in time and location.[9]

Around the same time, Smythies approached fellow SPR member and Oxford philosophy professor H. H. Price, who on February 9, 1952, had written to him: "I hope that you will not give up on the use of the word mind altogether. It is a dangerous word, no doubt, but I think the dangers of a purely Behaviorist theory of human personality are even greater." Before he left for Canada, Smythies invited Price to participate in a mescalin experiment. After Price took the substance, in London under Smythies's supervision, the professor wrote a thank-you note, adding, "I wish all my colleagues would try mescalin."[10]

In April of 1952 the first specific biochemical theory of schizophrenia made its debut in "Schizophrenia: A New Approach," by John Smythies and Humphry Osmond, which appeared in the respected psychiatric publication *Journal of Mental Science*.

Smythies had begun corresponding with influential individuals whose interests dovetailed with the mescalin-schizophrenia theory. It so happened that subjects frequently described experiencing similar visions, such as jewels and fantastical architecture and, because this seemed consistent with Dr. Carl Gustav Jung's theory of a collective unconscious, Smythies also wrote to Jung. The professor was very interested, as he wrote to Smythies in February of 1952, "I welcome unreservedly your idea of the Platonic *mundus* archetypes becoming visualized under the influence of mescalin."[11]

Smythies also sent Jung a copy of his paper "Extension of Mind," in which he described his conception of mind in a space with more than three dimensions. He also sent a copy to Aldous Huxley and, in a letter written November 25, 1952, Huxley replied. He reminded Smythies that, from the perspective of the philosophy of India, mind is regarded as extended as a subtle body in physical space. Huxley agreed that the usual notion of time-space was a problem when talking about communication between minds. Huxley added, "All this is avoided when the extension in space is at right angles to physical space."[12]

Shortly before he left London to join Osmond in Canada, Smythies accepted an invitation to spend an afternoon with Professor Jung, who had read Smythies and Osmond's paper on the mescalin visions with great interest. At the venerable man's home in Küsnacht near Zurich, the young guest enjoyed a generous tea and far-ranging talk in the professor's book-lined study, with its floor-to-ceiling shelves filled with tomes on mythology. Smythies recalls that at one point he asked Jung if he had ever tried mescalin. "He replied that he didn't need to," Smythies said. "He had lived in that world for much of his life."[13]

Not long after the meeting with Jung, John Smythies and his wife, Vanna, joined the Osmonds and moved to Weyburn, Saskatchewan. The move to Canada brought major changes in the lives of both couples. They had left behind a London mired in a postwar economy where basic foods such as meat, fats, sugar and sweets, bread, and even tea were still rationed, but this austerity was not the case in Canada.

Despite this benefit, and to Jane Osmond's dismay, Humphry's career had taken them to a place nearly devoid of cultural amenities in a landscape of bleak prairies. Weyburn lacked opportunities for ambling garden walks like those they had enjoyed but recently left behind in England. There was plenty of dismay for her husband, too, for Weyburn was understaffed and the two psychiatrists faced appalling conditions. Osmond, who described it as an Augean stable of perpetual filth, found upon his arrival one ward entirely devoid of furniture housing some eighty naked patients with severe intellectual disabilities, a ward where there were no toilet facilities other than a hole in the floor. As Smythies recalled, "Every morning the attendants would hose down the patients and the floor with powerful jets of water. I do not think that even Belsen [the Nazi concentration camp] had anything like that."[14]

Setting the hoped-for experiments aside, Osmond and Smythies had mountains of practical problems to solve. Soon Osmond was promoted from deputy to clinical director. In his new role he set about closing that horrific ward and providing patients with clothes, better sanitation facilities, and beds. Within a few years, his efforts would be acknowledged when Weyburn was named the most improved hospital in North America.

Dad's hand study mushroomed, and I couldn't wait for the 1952 Hobby Show. Aiming for the scientifically respectable sample of one thousand hands, he stepped up the collecting process in June with an exhibition booth at the Long Beach Convention Center, a WPA-era structure with exterior murals and a curving facade.

During the course of that weekend he'd hoped to add at least one hundred additional photographs. His display, with a sign reading "The Hand," had a disclaimer stating in large letters: "This is *not* palmistry." A chart explained how this was a scientific study for measurable correspondence between hand features and human character traits.

I was assigned the job of writing information on file cards and fetching coffee and sandwiches, anything to justify coming along. Our exhibit sign also challenged passersby to fill out a questionnaire, asking them to "Guess by number which hand belongs to the musician, the stutterer, the identical twin, the dentist, the successful insurance salesman." I sat in a folding chair, where above and around me hung numbered, oversized photographs of hands. The images appeared in reverse, the contours and lines of the palms and the whorls of the fingertips in white against black, the grain of the skin all sharply visible to the naked eye. The image was so sharp you could almost touch the padded part of the palm, one of the dimensions whose thickness Dad measured with a caliper. My father had sent a sample photograph to Harold Cummins, a Tulane University professor and expert on anatomy and dermatoglyphics—the study of the patterns of the skin. "The photograph that you enclose represents a beautiful technic [sic] and I congratulate you on the method," Cummins said in his reply, adding that he would be happy to learn about the results.[15]

At the hobby show, after we had set up our booth and the auditorium doors opened, people poured in. A surprising number of them began standing in our line. One man walked past and ignoring the sign said sarcastically, "This isn't a hobby, it's palmistry," but despite him our line soon stretched past other exhibitors and dog-legged at the end of the aisle, a queue of curious people, some of them skeptical-but-hoping-to-see-mysteries-revealed, most of whom towered over me as I sat in our booth on my folding chair.

My father looked confident that day and so did his signs, lettered with the precision of an engineer who quantifies, measures, knows what he is talking about. At six feet tall and with broad shoulders he looked professional in his white short-sleeved shirt. At his neck, a black bolo tie, his face tan from his recent desert outings.

A man stopped and introduced himself as a photographer and leaned in closer, asking my father how he captured such a fine degree of detail. My father said he'd be glad to explain it, for he liked to tell the story of how it took two years and half-a-dozen failures before one final adjustment of his lenses and lights and the horizontal hand-position ledge all came together as a purpose-built camera. He didn't mention me, his assistant, and the tiny darkroom in the living room coat closet where he developed prints in liquid-filled metal trays under a small red light, and where I joined him and listened to sloshing developer fluid, waited for faint images to burn in when the chemicals did their work, watched him fish wet prints from the tray and clip them to his darkroom clothesline. Sometimes our detritus in the closet exasperated my poor mother, who patiently picked up a debris trail of silver-lined Kodak paper packets and aired the house to drive away the tang of developer fluid, a smell I loved.

She had told me my father's hobby was unique, and on this day our exhibit was more popular than those in the categories of coin collecting or rock collecting, magic tricks or model trains. Though Dad had hoped to collect at least one hundred new images, his meticulous photographic technique proved too unwieldy for an impatient line of people, so we ended up mainly taking measurements. Even so, our exhibit attracted more interest than any other subject in the convention center, where I noticed that most people attending the Hobby Show were men, and most of them brought sons around my age, eight to ten being the prime age for hobbies.

In postwar America, where the theories of Freud, Watson, and Skinner held sway, the clinical mental health community struggled with harsh facts behind secured institutional walls. It was one thing to treat an affluent urban professional undergoing psychoanalysis and quite another to manage the problem of thousands of patients overflowing the public psychiatric hospitals of North America. When it came to schizophrenia, the five-hundred-pound gorilla of psychoses, there was no explanation, much less a method of early detection for this disease that most often struck young people, usually males in their teens or early twenties, as is still the case today.

Then, too, overcrowded mental hospitals typified a postwar period. Many patients had suffered from what in World War I was first called "shell shock." As Joseph Engel points out in *American Therapy*, when the draft call-up began after December 7, 1941, a surprising number of arrivals at induction centers were identified as already having mental health problems. From January 1942 to December 1945, 12 percent of inductees were rejected for neuropsychiatric reasons; by the end of the war, 40 percent of medical discharges had been for psychiatric reasons.[16]

Physician William P. Menninger, chief psychiatrist of the Armed

Forces, was tasked with the problem of mental health in the military during wartime. He also cofounded the Menninger Psychiatry Clinic in Topeka, Kansas, with his brother Karl, who will resurface in this story slightly later, in the mid-1950s.

Switching for a moment from the topic of mental illness to mental health, it turns out that a much-adopted psychological tool in the first half of the twentieth century was a personality inventory based on typologies. The starting point was C. G. Jung's theories in his book *Psychological Types*, published in English in 1923; this led to a practical application by Catherine Cook Briggs and her daughter Isabel Briggs of a personality inventory created during wartime in 1942, initially to help women fit into the industrial war effort. This later became the Myers-Briggs Type Indicator in 1956, a psychometric questionnaire used in career and educational testing and counseling.

In a way, the hand study resembled a typology but not a personality inventory, because it did not start with a questionnaire based on behavior. But there was great interest in such approaches in the postwar 1950s because fitting a man (it was usually a man) to the right job and doing it expediently was potentially more productive than making a bad match. That had been the purpose in wartime of my father's step-up-problems workbook—to quickly identify talent and sluice out the less adroit.

In the postwar years at North American Aviation where my father now worked, the personnel director was interested in any technique that could improve the quality of personnel selection. Perhaps because his fellow engineers and draftsmen were so convenient, the first phase of my father's study had focused on photographs and measurements of their hands. Some came to our house to be photographed and sometimes he brought his camera equipment to their homes. Some subjects were curious about typologies and wanted to know more, and some probably thought this was just their coworker Howard Thrasher's eccentric and time-wasting hobby.

After he had collected several dozen photographs of the hands of current employees, my father analyzed the data and came up with the hypothesis that a hand with a wide palm and long fingers indicated an individual who had more potential at a drafting board. By way of proving his point, he met with the North American Aviation personnel director, who had agreed to take a look at Howard's findings. On that day he brought the finished prints to his office and arranged them in three stacks on the desk. One consisted of current employees who had hands with wide palms and long fingers. The other two stacks represented examples of narrow palms or short fingers.

The photographs were arranged face down so the personnel director saw only names written on the back. He immediately singled out the stack with names of employees who were regarded as the most productive draftsmen. Howard turned them over, like the revealing moment in a card game, and when he did (or so I've been told) it was clear that the wide-palmed–long-fingered group in this stack stood out from the rest.

According to my father, the personnel director was sufficiently convinced to try this technique visually, as a supplementary hiring aid, but when rumors began circulating that something strangely like palmistry was going on in the department, the director dropped the idea and Dad was told to leave any hint of his controversial hobby at home.

But it turned out that he and his hand photographs were quite welcome elsewhere. The state mental health agency was under pressure, because, after the war, California's mental hospitals were bulging at the proverbial seams. Patients in mental hospitals seemed like a worthwhile category for study, and after Dad contacted officials in Sacramento he received a reply dated September 11, 1952, from the State of California Department of Mental Hygiene. It granted him permission to conduct his study with "cooperation but not at taxpayers' expense." It seems officials were willing to open the door for the long-shot possibility of a new typology, maybe even a diagnostic aid. This is how the doors opened for my father, the amateur researcher, who gained access to Norwalk State Hospital, and later to Pacific Colony in Pomona, where he would be allowed to photograph and study the hands of schizophrenic patients.

In 1952 the *Hibbert Journal*, a review of religion, theology, and philosophy, published an article by John Smythies and Humphry Osmond about the current state of psychological medicine and its resistance to a biological approach.[17] It mentioned how a psychoactive compound such as mescalin induces a derangement resembling schizophrenia but at other times seems to induce what is known in religious literature as a visionary state. This was a compelling dual concept, or so it must have seemed to Huxley, whose latest book, *The Devils of Loudun*, dealt with a seventeenth-century case of alleged diabolical possession of a dozen Ursuline nuns, and the attempt of one Father Surin to save them, using prescribed rituals of exorcism. Father Surin subsequently suffered from hallucinations himself, though he carried on with his clerical life. Huxley saw the retelling of this historical incident as a chance to explore strange aspects of the mind.

The *Hibbert Journal* article probably reached the Huxleys' mailbox in Los Angeles on a balmy late autumn day in contrast to Saskatchewan, where the hospital for the mentally ill would have faced a changing season's early chill.

Weyburn had turned out to be a horror show of mismanagement at the time Osmond arrived as its assistant director,[18] but sometimes walking into a bad situation can make you look good. In this case, once the director was dismissed and Osmond inherited the top job, he gained greater freedom to do research, although at the cost of being shackled with compounded responsibility. The Weyburn research team soon expanded, with Dr. Abram Hoffer, an MD with a doctorate in biochemistry, and with the arrival of Dr. John Smythies.

The three men set about working with mescalin and other substances that simulated schizophrenia, searching for a possible biochemical basis for the disease. Osmond and Smythies, who had begun pursuing this research in London, expanded on the theory they had set out in their first copublished article.

Osmond had been director of Weyburn for a few months when, in March of 1953, a book arrived in the hospital mailroom addressed jointly to Doctors Osmond and Smythies. The rectangular package weighed slightly over one pound; I can vouch for its weight because the black book with no dust jacket that I found in our garage was also a first edition of *The Devils of Loudun*, Huxley's retelling of a documented seventeenth-century incident of what some would call group psychosis.

The return address on the package in the mailroom read 740 North Kings Road, Los Angeles 46. Inside the package was a note from Huxley expressing appreciation for the article. Clearly Aldous was intrigued by the idea of mescalin, mystics, and madness.

Weyburn Hospital sheltered, if you could call it that, more than 1,800 of the gravely mentally ill. Opened in 1920 and spread over eighty acres, it had three large wings converging at a main entrance where, in a photograph, a brass handrail appears to hinge the facade to its circular drive. Dr. Osmond at age thirty-six was now on a steady yet conflicted upswing in his medical career. Osmond and Smythies' controversial research, along with Osmond's appetite for challenging the establishment, would propel him into a circle of fame he would one day share with the famous author of the book newly arrived in the mailroom.

But at this point the hospital's new director had to keep his feet on the ground. He was experimenting with ways to improve the abysmal living conditions as well as the prognosis of the patients who occupied the hulking brick building with its chronically foul-smelling rooms, where chemical cleaners didn't stand a chance against the surge and stench of human waste. In keeping with the prevailing institutional theory, severely ill mental patients were not expected to improve or ever return to the outside world. Today one would call them warehoused. Osmond, driven to solve problems, wanted these abhorrent conditions changed.

The other branch of Osmond's career, and hence his conflict, was riskier and more exciting, and to this pursuit he devoted every moment he could spare. His quest, in conjunction with Smythies, was to find, if it existed, a naturally present human substance linked to the incurable disease called schizophrenia. Victims of this disease ranged upward in severity of symptoms from a mild hallucinatory fog to a blinding, blaring world of hell.

When he had first arrived at Weyburn, Osmond saw the nightmarish confines of the worst patients in that one ward, or as he would later recall, "The unfortunate patients with their own perceptual distortions would crouch, terrified, against the walls. . . . The patients were particularly miserable in their huge dormitories and dayrooms; sometimes a hundred of them—unclothed—would be milling around, as in a concentration camp."[19]

I would see a photographic glimpse of this horror when I visited Osmond's daughter, the day we were sorting through boxes of her father's papers and memorabilia. I opened one folder and gasped at what looked like Auschwitz on the day it was found by Allied troops, but these were Weyburn patients, with protruding ribs, matted hair, mouths twisted in anguish, standing huddled against stained walls. When I saw this photograph I had a sense of what Osmond must have faced when he arrived as deputy director of the hospital. Within half a dozen years he would use an approach that came to be called socio-architecture to improve patient settings. (Osmond coined this term while working with Canadian architect Kyo Izumi to reconfigure the wards and dayrooms of the hospital.) The earlier photograph was used as evidence of his successful campaign to bring about such a change.

How I came to have this shocking photograph in my hands, and how I ended up one weekend at Humphry Osmond's former home, brings me to the day I met his erstwhile research partner.

It was January 22, 2007, and I sat in the office of John Smythies, MD, director of the Center for Brain and Cognition at the University of California, San Diego. The window formed a frame behind his head, and, beyond the glass, eucalyptus leaves shivered in the afternoon breeze.

The professor leaned back in his chair. He was tall and thin, blue-eyed and pale of complexion and wearing a droll smile. He was bemused, I imagine, to see three acolytes (visiting scholar Bill Rosar, Bill's friend Alan, and me) lapping up his words.

On that winter day, before I knew Humphry Osmond even had a daughter, I was about to begin an interview that had been arranged by my friend Bill, a Research Fellow in the Center for Brain and Cognition at UCSD. Trying to plug large gaps in my research, I began by asking about events that took place fifty years ago involving Humphry Osmond and Aldous Huxley, mescalin and LSD. At some point Smythies said something along the line of "No doubt you have spoken to Fee." I fished around in my brain and wondered whether Fee was a he or a she and I actually thought it sounded more like a noun, a tax, a tariff. "I'm sorry," I said, "I'm not sure who this is."

He informed me that Fee was short for Euphemia and that I should contact her right away. He apparently had her e-mail address at hand and in a flash pulled up his mail program, followed by a few clicks, and told me he had just sent her direct contact information to my e-mail. For someone in his eighties, he showed no sign of slowing down.

Maybe the professor thought I was wasting his time, but if so then I was grateful to him for concealing it, considering how clueless I'd been when I arrived at his door a half hour before this. For example, soon I learned that Huxley and Osmond had been involved in a project called Outsight, and I learned about a medium named

Eileen Garrett who held international parapsychology conferences both Huxley and Osmond attended. I had risked the embarrassment of fools-rush-in by assuming I had only one chance to meet Professor Smythies, so I clumsily crashed ahead unprepared, not guessing in how many new directions this meeting would take me.

A few days later, back in my home office, I planned a list of questions to ask Humphry Osmond's daughter. With more time to do my homework, I caught up on the highlights of her father's life and career, among other things finding out he was born in 1917, the same year as my father. The next step was to introduce myself by e-mail and tell her about my plan for the book, though that plan was changing daily. After a few electronic replies we arranged a time for a telephone call. As a journalist, I knew interviewing a subject by phone could draw out more surprises than even the most detailed e-mail query, and our talk would be no exception.

As I entered her number, I anticipated hearing stories about growing up, about her life in the looming shadow of Weyburn mental hospital, perhaps a few insights about her father's relationship with Aldous, but when she answered, my first impression of her voice was to think *this seems odd*, for most Canadians do not have such a distinctly clipped British accent. Then Fee explained the reason for hers: She was born in Weyburn, but her mother brought the children back to England while Humphry remained behind to continue his work, which meant that by the late 1950s she and her mother, sister, and a brother born after Fee were residing in Surrey at Onet Cottage, her traditional family home.[20]

But I was puzzled, because this must have been in the peak period of Osmond's work with mescalin and schizophrenia. "My father would come back about twice a year," she said. "It was an odd arrangement." No, she told me, her parents did not divorce, nor was there enmity between them. The children, I figured, must have had far greater educational and cultural advantages across the pond. Then I asked, "Do you by chance have documents such as letters exchanged between your father and Aldous, or any mementoes of their friendship that you might be able to describe to me?"

"I have a large binder of my father's, filled with their correspondence. There are maybe one hundred letters." She hesitated for a moment then added, "The easiest thing, I suppose, would be to send the whole binder to you. You could take your time and do your research, then send them back to me."

I thanked her profusely, thinking of the source material I'd soon have in hand, picturing photocopies or maybe copies of carbons. After all, we're talking about documents from the 1950s, not digitized. And a few days later the FedEx package arrived. I opened the flap. Inside I saw a black binder, just as she had described it. When I slid it out of the shipping box, the smooth composite material of the cover feeling cool to my touch, I saw a label, "Huxley correspondence 1953–1963," opened the cover, and found not only Osmond's carbon copies of those he sent Aldous but also scores of *original* letters from Huxley—some typewritten with his strikeouts, some written by hand on postcards from different destinations around the world, some on the letterhead of legendary trans-Atlantic ocean liners.

I flipped through them in amazement, gazing at this trove with Huxley's scrawl and his signature on page after page, and it is no exaggeration to say that I was shaking with excitement, as would be any biographer. Halfway through the binder I came upon one of the famous letters in which Osmond and Huxley wrote their dueling rhymes, each man making his case for a different name to call what at the time was a growing family of visionary drugs.

Soon after FedEx dropped off the box I cleared my calendar for two days to read through the sequence of letters. I pored over them, reading for hours at a stretch, mesmerized by the dialogue between the two men. While I was in the midst of reading, my friend Rudy, a former rare-book dealer and someone I hadn't spoken to for months, happened to call, and I told him about the letters. "She barely knows you and she loaned you the Huxley *originals*?" He sounded upset. "Those are very valuable. Why would she take such a chance?"

I thought of Fee telling me how, when her father crossed the Atlantic for one of his twice-yearly visits with the family, "all sorts of interesting people came to Onet Cottage." I thought of my parents' séances and how a home can become a magnet for people of like mind. "I guess she trusts me," I told Rudy. "We have a lot in common." Our dads were enmeshed in experiments that coincided with our childhoods. Humphry Osmond and Howard Thrasher both became involved with Aldous Huxley, and both fathers believed their findings would benefit all humankind.

CHAPTER 5

STRANDED IN THE STUDEBAKER

I n 2006 I drove to Norwalk, California, to see what is now called Metropolitan Hospital. After a turn onto Norwalk Boulevard, past a stretch of small homes, the city bustle receded and I felt a prickle on the back of my neck. First I saw a fence, then the towering and now-shuttered administration building of what in 1952 was known as Norwalk State Hospital.

Half a century before, this place had sat on 167 sprawling, wooded acres, and though it is slightly smaller in area today and has newer buildings mixed with the old it still seems a relic, an eerie rustic retreat juxtaposed against warehouses and thrumming freeways. By then it was late afternoon and three television traffic helicopters were whirring overhead. I slowed down to stare at the old building and the closed-off driveway entrance that I am quite sure was the same one my father and I drove through one Saturday long ago.

In 1952, when Dad received permission to photograph the hands of patients at the hospitals in Norwalk and Pomona, his schizophrenic subjects were not as severely psychotic as those in the pictures taken at Weyburn, not by a long shot. In fact, one of the Norwalk patients so cleverly passed for what were then called "normals" that on that occasion it even made my dad forget that I was waiting for him in the car.

"Walking into the ward you'd think you were walking into a ward of normal people," Dad told me as we sat on the patio one day, "but

I was there so many times that I finally got to see this crazy streak in all of them."

As a kid, that day I had come along for one of those Saturday sessions, probably because my mother and grandmother were busy at some appointment. Dad and I set out for Norwalk, about a dozen miles northeast of our home. His collection had reached several hundred hands by this time, and statistical weight urged him toward one thousand like a brimming cart rolling downhill. We knew unaccompanied children were not allowed to wait alone in the lobby so my father moved the Studebaker to a space close to the lobby entrance and I made myself cozy in the car. It's not something you'd be likely to do today, leaving a kid alone in a car, but this was the 1950s, and I'd brought along books and lunch. That morning, before leaving for her appointment, my mother had prepared sandwiches. By now it was almost noon.

"Are you okay here?" my father asked. Yes, I told him; I had planned to reread my favorite book until he came back, and then I asked when that would be. About three p.m. at the latest, he said, as I recall, and gave me his wristwatch, which, chances are, I set on the dashboard. He stood next to the car window, bending down to me at eye level. His tan, wiry arms supported the base of the black plywood box housing his photographic equipment, its spider-leg strobe extensions folded inside.

"There's a bathroom in the lobby," he said. "If you don't see it, just ask."

Eventually I unwrapped the crackling wax paper and munched on my sandwich. Reading in the backseat, my legs stretched out on the fabric upholstery of the Studebaker, I became engrossed in my book and lost track of time. And that's what was happening inside the hospital, too—Dad was losing track of time. By late afternoon he had taken more photographs than he expected. He ran into some technical difficulties so he took those shots over again. Without my mother or me along as his assistant and recording his own measurements each time, every part of the process took longer than he'd planned.

Finally, he began to pack away the equipment inside its self-contained box. The patients whose hands he had photographed and measured for palm and finger dimensions had returned to their rooms or to a common area in that wing of the hospital. He later told me why he had lingered.

A woman named Veronica struck up a conversation. She was attractive, neatly coiffed, with carefully applied makeup and dressed in a fresh pastel smock. He assumed she was one of the technicians. He wasn't particularly susceptible to feminine wiles, except this woman engaged him in a technical conversation and I know how that could grab him. "She asked what I was doing and seemed genuinely interested," he said. "She asked about my equipment, the kind of emulsion, my film and paper. She knew all about the developing process, maybe more than I did."

Wrapped up in this conversation, he forgot all about me. Plus, he was not wearing his wristwatch, which he'd left behind and was now on the dashboard. Waiting inside the car, I became frustrated when the sun started to drop below the horizon and I had to twist awkwardly to catch the light in order to read. Soon dusk was turning into night and I was alone. The parking lot was almost empty and what had been annoyance turned into fear.

Inside, my father was still talking to the female photographic expert until mid- sentence, without warning, her face twisted into a mask and she began shrieking at some imaginary foe, shocking Dad into realizing that he had not been conversing with a technician after all. Soon he had packed up his gear and scurried out and was back behind the wheel of the Studebaker, apologizing repeatedly while he drove us home.

Afterward, and from then on without bringing me along, Dad

conducted other sessions at Norwalk State Hospital and afterward at the Pomona facility, but none of them had as much drama as that one day—that is, not until he sat down with one of the hospital directors to talk about his findings.

Not long after the Saturday when I accompanied Dad to Norwalk State Hospital, I remember I stayed home with a cold and passed the time with a rainy-day workbook. These had puzzles like connect the dots, and I was pushing my pencil through a maze when at some point my father stuck his head through my bedroom door. He was carrying one of his experiments to try on his glad-to-be-a-guinea-pig daughter. He held out a sheet of graph paper he had inked with a black nest of lines as if scribbled by a two-year-old, except the artist was my dad. My pencil sat idle. I saw nothing to connect or decipher. "What do I do with this?" I asked.

"It's like the games in your workbook," he replied. "Look closely and you'll find the hidden cow."

I guess my father had been dabbling with decrypting something the Gestalt psychologists called "hidden figures." But when I turned the page every which way I saw only shapeless scratchings or what looked like a disgusting tangle not unlike a mass of hair stuck in a bathroom drain—but maybe a similarly grotesque thought set loose some free-association process and suddenly the hidden creature appeared. It was as if the rest of the tangled lines receded and Bossie stood out in relief. I saw the contour of her big head and muzzle, the bony back above and bulging udder below the hind legs, as if cow was mooing at me, *Hey, here I am.*

The discovery of a possible diagnostic indicator for schizophrenia in the patterns in the hand seems to have transpired rather like finding a cow in the tangle of lines. One weekend in late summer of 1952, as close as I can reconstruct the date, the director from Norwalk State Hospital and my father met to discuss my father's findings. They didn't meet in a cluttered office at the mental hospital, Dad tells me, but in the doctor's spacious apartment. This was largely in order to have adequate floor space to take the next step.

"On this side I'll lay out photographs of two dozen normals," my father recalled saying, as he proceeded to arrange dozens of life-sized prints on the floor. By normals he meant the control group, samples he had collected randomly from people with no particular affliction, and probably many of them were from the Hobby Show.

I picture a midday sun filtering through the apartment's big front window, where on the other side of the room Dad set out another two dozen or so photographs. These were patients diagnosed with schizophrenia. He added the last few prints, then said, "Now, let's see what we have."

I am told the two men circled the enlargements, pausing to lean over them, coming in closer and bending down on their knees, their movements punctuated by an occasional murmur. Maybe after a while they became bored, even groggy, the doctor questioning the wisdom of spending his Saturday afternoon this way and maybe thinking, *Well, Howard, you have gone to a whole lot of trouble for nothing.*

Dad recalled drawing himself up from his crouched position on the floor and saying words of encouragement. *If we look hard enough I have a feeling we'll see a difference, some anomaly.*

They might have stopped for coffee or a smoke for the sake of alertness, the eyes of both men reaching a point of fatigue and certainly the point of eyes glazing over with boredom, but sometimes such a low point is precisely when a breakthrough occurs, when reason flags and ordinary focus blurs and the ground recedes and patterns take over and maybe a newer and truer shape emerges from the tangle of lines.

"Wait a minute," the doctor said, turning toward the examples on one side then the other. "I see something different about the right hands in this group. Look at that little area of the grain pattern under the index finger. It's like a triangle."

My father went down on all fours as the doctor traced the pattern with one finger. "Yes, I see it!" He remembered the doctor muttering something about this being beyond chance. He remembered the doctor saying, not with any joy, *what have we found?* because, if schizophrenia turned out to be quite literally ingrained, then a search for a cure would be pointless. They scrutinized both sets of photographs. None of the hands of the so-called normals had this pattern, not one.

On January 28, 2002, five decades later, my father and I planned to talk about the mark my father found, the indicative "cup of grain" that he and the doctor had seen in the hands of schizophrenic patients. Dad asked me to bring along a certain box of photographs from the garage.

We were meeting at Dad's apartment, about two miles north of my mother's bungalow. The day before, he'd called to say he didn't feel up to driving his delivery-van camper. Despite his health food diet and meditation and his long walks, he felt under par for some reason. But he was eighty-two, after all, and I knew someday I would lose him and maybe this would happen not long after I'd found him. When I arrived at his place I saw that his face looked drawn as he sorted through the stack of photographs.

"This is a good example," he said, pointing to the area just below the index finger. "When the doctor and I spread the pictures on the floor of his office that day, we both had the same idea. There's a story here."

A truck rumbled past while he drew a diagram and set the piece of paper next to the photograph. "I have coined the term "high-low" for these hands. Here is the starting point of the triangle." I looked closely and, sure enough, the blow-up revealed a shape just like the diagram. I could see it below the index and the middle finger.

"That is the cup of grain. Where the grain pattern comes to a point is our starting point," he says. "Where it goes from there is what we look for. In some hands the grain pattern flows up or down. It is a common feature in all hands, and it is found in both hands of any individual."

He turned the photograph of the hand around so it faced me. "In most people it's very easy to trace the little triangle, especially if you put white powder on it so you can see it better. You've got to have a good magnifying glass. You take a pen and mark the line, and if you've lost it you start over. If you find that on the right hand it is coming out and going up over here"—he traced it with his fingertip—"and on the person's other hand, the left, it's coming out and going down over there, that is the high-low pattern."

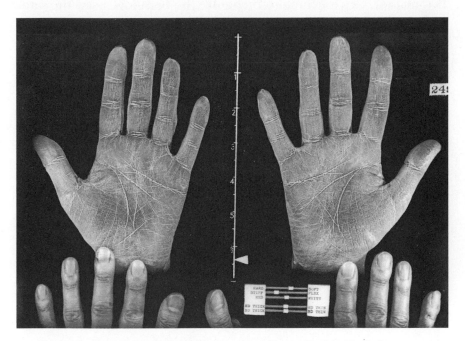

Figure 1. Analytical photo of the hands of Aldous Huxley. Photo by Howard Thrasher. Property of the author.

That was the one side up–one side down pattern that appeared to indicate schizophrenia in an individual. "In hundreds of hands," he said, "I've never seen the reverse: a pattern with the grain in the left hand going up and the pattern in the right hand going down."

"Then how does it look in a so-called normal person?"

"When the left and right hand are the same, I'd say those people are rarely found in mental hospitals."

"So is it just this one mark?" I asked.

"I think the grain structure tells a bigger story and it hasn't been explored. I've seen other things, like disconnects in the grain that I think indicate a troubled person. I saw enough to know one thing—there is a code in the hand." Then he added something that took me aback. "Schizophrenics were not the only ones who had this marker. Once I came to recognize the high-low, I saw the same pattern in another group. The photographs I took at the convention of psychics."

CHAPTER 6

TUESDAY NIGHTS ON NORTH KINGS ROAD

I'd say 1952 was a watershed year in the decade of midcentury modern. By Christmastime television sets were already in half of the nation's homes, and if a black-and-white screen measured twenty-one inches it aroused considerable envy.

Within a few weeks, when 1953 rolled around, Walter Cronkite would host his weekly show *You Are There*, a CBS drama reenacting highlights of American and world history—call it an early take on what we now call docudrama.

In the spirit of Cronkite and his television reporters (who, oddly enough, interviewed costumed historical characters while wearing modern-day suits), I would like to introduce some of the guests who attended Aldous Huxley's Tuesday night salons circa 1952. This first reenactment takes place at the corner of Beverly Drive and La Cienega Boulevard in what is now West Hollywood, where the action unfolds inside an Owl Rexall pharmacy deserving of its slogan, "The World's Biggest Drugstore."

The drugstore cafe has become a gathering place for guests who first dine here, then regroup at the Huxleys' home on nearby North Kings Road. Tonight the diners cluster in twos and threes, in booths or perched on lunch-counter stools, a dozen or so men and women clad in early 1950s attire. The women are sheathed in slim pencil skirts or wear black toreador pants, the men in gabardine slacks wearing sport coats over white long-sleeved shirts with ties.

One of the diners is Hereward Carrington, a longtime member of the Society for Psychical Research, and hand number 246 in my father's photographic collection. He is known for his investigations of anomalous phenomena and sometimes debunks psychics, but has been impressed by a renowned medium named Eileen Garrett.

I imagine a dialog taking place in the diner:

"I am far from confident that Rhine will prove the existence of ESP through his technique of accumulating data," Carrington might say, referring to J. B. Rhine, founder of the Parapsychology Laboratory at Duke University "But if he does, then perhaps funding will allow us to explore the promising subject of precognition." Huxley is quite familiar with Carrington's theory of precognition and has recommended it to Rhine.[1]

A woman with her hair twisted into a chignon takes issue. "If you want to see progress in mind-to-mind communication then I assure you it won't happen with J. B.'s damned boring Zener cards. His technique lacks the juice, the crisis-stimulation likely to trigger clairvoyance."

Carrington might counter: "And how would you go about creating a personal crisis in a laboratory setting?"

"I think what we need," the chignon woman says icily, "is a new approach. Something more radical."

The pre-salon group is gathered at the World's Biggest Drugstore tonight because Maria Huxley, though always gracious and supportive of her husband Aldous's projects and friends, does not provide Tuesday dinners for a fluctuating roster of guests. Some of these people will be invited back, others she will never see again. Apart from its convenience for these dinners, the proximity of the mammoth drugstore serves two other purposes, one being a shopping resource for Maria.

But on some days the Owl Rexall doubles as an amusement park for Aldous, who despite his impaired eyesight always delights in taking ironic jaunts through the aisles. He peers through his spyglass at the bizarre juxtaposition of products, some, like Mr. Potato Head ("the funny-face kit"), mainly targeted to youngsters, located an aisle away from cone-nosed undergarments ("I dreamed I was bewitching in my Maidenform bra"), an aisle away from *Mad* magazine in the periodicals section, where comics are adjacent to art books—all this reflecting the crazy quilt of the American consumer's appetite.

Aldous doesn't know it at this point in 1952, but next spring this cabinet of curiosities will play a role in an experiment far more radical than J. B. Rhine's Zener cards.

Meanwhile on the corner of Beverly and La Cienega, the meal is ending at the Rexall café. The diners, some of whom have met each other for the first time, pay the cashier for their individual tabs, then pile into Studebakers and Chevies and Hudsons and drive about a mile to the Huxleys' home, where they park along the tree-lined street.

Aldous and Maria welcome their guests at the door. Wisps of conversation stirred up at the café drift through the entry hall and trail the speakers into the music room. People alight on a sofa or one of the mismatched chairs, for Maria is a rather haphazard decorator. Long ago she decided décor wasn't worth too much trouble in their house because Aldous wouldn't notice and ideas were mainly what mattered here anyway.

On any given Tuesday night those ideas might turn into demonstrations or experiments. There might be a séance or age-regression hypnosis, or perhaps an account by someone involved in remote healing. Talk sometimes circles back to the lurking possibility, hinted at in the demonstrations, of personal survival after death. It all depends on the mix, the guests in the Huxleys' music room on a particular night.

And some nights the topic is set by one featured guest. On this occasion it is Howard Thrasher, my father, who will talk about the project he calls the Hand. He has brought with him a set of black-and-white blow-ups that Aldous and Maria have seen before but the others who are present have not, so he passes these examples around. He opens an odd carrying case, pulls out a large camera, and extends a pair of strobe-light arms so the assembled equipment looks rather like a predatory insect. Once the camera is in place, he will begin the rather messy procedure he uses to capture hand images of some of tonight's guests.

Before he does this, however, he will explain some of his findings. According to his research, he tells them, the palm and fingers—but especially what he calls the "grain pattern" found on the skin when its minute details are photographically revealed—offer insights into personality traits and proclivities of an individual. The grain pattern also indicates the presence of severe mental illness. From this study, he believes, he can show how the hand is a mirror of the mind.

Who were these Tuesday night people? Psychics and healers and hypnotists, to name a few. Some lived in Los Angeles or elsewhere around Southern California, while others were passing through town and came from thousands of miles away. Some were invited after Aldous noticed a name mentioned in an interesting article, the kind of articles Maria (perhaps with an "I think you'll like this, Aldous") read to him aloud. And sometimes he relied on referrals from friends like Gerald Heard.

Aldous extended invitations to individuals who spawned fringe-of-science ideas likely to arouse controversy, lead in new directions, challenge the laws of nature, and sometimes inspire Aldous to write an essay and maybe even a book. Aldous, his nephew Francis Huxley once told biographer David King Dunaway, "liked the company of large minds with obsessions."[2]

Though Aldous wrote several thousand letters in his lifetime, the mix of ideas at the Tuesday salons must have generated a strange energy. Aldous was an involved host, who deftly probed his guests, or, as my father bluntly put it, "He picked your brain." Aldous had been a journalist, and journalists usually follow leads, except that in the salons in the house on North Kings Road, Huxley's leads came to him.

Dad was impressed by the way Aldous grappled with complex ideas. "It might take me, or someone else in the group, five minutes to express an idea," Dad said, "and Aldous would sum it up in one sentence."

Though Dad was a confident speaker in his professional field of aircraft design, when talking about the Hand he was building an untested scaffold of concepts. When something sparked Huxley's interest, he would process it through his encyclopedic knowledge of science and literature and history and ancient and modern languages until a fitting phrase or workable metaphor emerged. "From Aldous I learned how stupid I am," Dad said, laughing, but with a shake of his head as he recalled the experience. "Sometimes he would say, 'Do you mean this . . . or do you mean *this*?'" My father admitted to feeling conflicted. With great effort he had excavated the raw material and hauled it up to a conscious level, after which Aldous refined the idea with his great smelter of a synthesizing mind.

Apart from the Tuesday evenings, sometimes an invitation was extended for dinner at the Huxleys', which was another way for Aldous to appraise a large idea and the originator's obsessive mind. In the early 1950s, when Huxley was plumbing the depths of the human psyche, at the time he was writing *The Devils of Loudun*, one of those who came to dinner was science fiction writer L. Ron Hubbard.

When Aldous invited Hubbard and his second wife, Sara, to dine at North Kings Road, it was around the time Hubbard's book *Dianetics: The Modern Science of Mental Health* was published. Huxley was curious about Hubbard's notion of identifying and erasing negative forgotten memories, the so-called reactive mental impressions that Hubbard called "engrams." Aldous is said to have undergone four sessions and realized that he had shut off certain memories of childhood.[3]

My parents were the Huxleys' dinner guests a year or so after that evening with the Hubbards, and imagining my parents dining at the Huxleys seems as odd as any weird product adjacency in the World's Biggest Drugstore. I am told that the dinner took place in the Huxleys' formal dining room, at the heavy wooden table that occasionally played a role in Tuesday night séances. On this occasion, the erudite, tweedy Aldous and the petite, stylish Maria, both in their fifties at the time, sat across from my housewife mother and engineer father, both in their thirties. I am told that my mother made Dad wear a sport coat and a tie, though he at first resisted.

Dad's discomfort in polite company points to the contrasting backgrounds of the two couples. The Huxleys shared interests spanning literature, fine art, and world travel, ever since their marriage in London in the 1920s. They had socialized with world leaders and counted among their close friends such luminaries as astronomer Edwin Hubble, who had changed our understanding of the universe by confirming the existence of galaxies beyond our Milky Way, and conductor and composer Igor Stravinsky, whose world renown stemmed from his *Le Sacre du printemps*, or *Rite of Spring*, and other compositions for Sergei Diaghilev's Ballets Russes.

Other differences between the two couples are too numerous to list, but one was their choice of automobiles. Over the years, the Huxleys had owned various vehicles—the most legendary being the scarlet Bugatti of their early married years, driven with precision by Maria. My father drove a pale-yellow Studebaker, and instead of an appreciation for fine art, my dad's connoisseurship involved tinkering with a bread delivery van converted into a camper he sometimes used for his remote desert getaways.

What Huxley and my parents *did* have in common was an interest in experiments. Dad's intense focus on his growing collection of hand photographs appealed to Aldous. That night when my parents were dinner guests, Maria brought out examples from her own collection of palm prints, an interest she had first taken up in the 1930s. She said she had made these impressions by rubbing ink on a subject's palm then pressing it onto a sheet of paper. Her collection included the hands of many dignitaries and celebrities the Huxleys had encountered in their world travels.

When I asked him about the Huxleys' house all these years later, Dad said he'd paid little attention to it. But his wife had taken in the surroundings that first night, and again when she returned on occasional Tuesdays. My mother noticed the hardwood floors, the statues and etchings and prints on the walls. She told me, looking back on her initial impression, "It was like visiting a museum in a private home."

That first evening she had occasion to walk past an open door and, being curious, she glanced inside. Immediately she knew this was Aldous's study, where he wrote his essays and articles and books. "The little room was serene and stark," she recalled, "but it had a beautiful atmosphere, spare, nothing to distract."

She sat across from Aldous at the dining room table, and it was obvious that he had impaired eyesight. He gazed intently at people and objects. What impressed her, she said, was "the way Aldous seemed to focus on everything, even more than if he'd had normal vision."

It is another Tuesday night on North Kings Road, and five new photographs will be added to my father's collection, numbers 280 through 284. I am at least able to give a sketch of three of the guests: Jean Dunn, hand number 280, devotee of a revered east Indian sage; psychologist Leslie LeCron, number 282, whose specialty is experimental hypnosis; and an unknown woman whose hand number is 283 and is identified by an index card that reads, "Roy Maypole's girlfriend."

The first subject, Jean Dunn, is an occasional visitor to Los Angeles. She extends her hands toward Howard and watches as he applies the white photo-enhancing mixture to her fingers and palm. She lives most of the year in Bombay, where she is a devotee of Sri Nisargadatta, also known as Nisargadatta Maharaj. Dunn is dedicating her life to transcribing and editing his talks.

Howard presses the shutter release. Jean, a trim woman with short, curly brown hair and glasses, leaves the music room to wash off the white emulsion and afterward steps outside for a smoke.

Next, my dad photographs the hands of psychologist and hypnotist Leslie LeCron, number 282, whose book *Experimental Hypnosis* was published by Macmillan around this same time, in 1952. Among his other published work will be an article three years later called "The Paranormal in Hypnosis."[4]

LeCron's specialties overlap three recurring Tuesday night topics, one being hypnosis applied in unorthodox healing. Another is a study of spontaneous mystical visions that at times occur during hypnosis—and he has found that these are remarkably similar when described by different individuals. Leslie is also a consultant for the Los Angeles Society for Psychical Research. Lately he has been experimenting with so-called age regression to explore the possibility of past lives and the survival of consciousness after death.

The next guest to have her hands photographed is the girlfriend of television producer Roy Maypole. Her hands are number 283. For some reason I picture her wearing 50s fashion-appropriate high heels, and if so then eyes are likely to follow her progress across the room to where my father takes her hands and smears on the chalky substance. She places her hands on the alignment ledge. He turns on the strobe lights. Merely by being here she is game to try something new.

Her date, Roy Maypole, probably stands to the side.[5] Apparently he has wangled an invitation by saying he wants to schedule a television segment about Huxley and his friends and their unconventional interests. I picture him shifting his feet, arms folded across his chest. You can see that he is skeptical. And when Maypole finally does produce a show on this general subject, or so my dad tells me, it will not be a discussion of pro and con but hard-angled ridicule.

After the hand photographs have been taken, everyone settles down for the evening's discussion, and, though I am speculating about the exchange of ideas that night, I think part of the conversation might have gone something like this:

"Most of you are familiar with Vedanta, the Western interpretation of Hindu religious thought," Aldous might begin. "I would like to ask Miss Dunn, who has traveled here from Bombay to please tell us something about Sri Nisargadatta. You are translating his writings, I understand."

In a voice made husky from cigarettes, Jean Dunn replies, "The heart of the maharaj's message is simple. Enlightenment requires no lengthy discipline, only an unceasing awareness of *I am.*"[6]

Someone else ventures, "I recall that St. John of the Cross writes about emptying the memory as being a great good."

Aldous might take up the idea of memories, recalling an essay he had written a while back for the Vedanta publication. More recently he had tried Dianetics exercises, with their odd probing of memories. "There is a disquietude in dwelling on our memories," he might recall from his own essay. "Disquietude and distractions constitute a formidable obstacle to any kind of spiritual advance."[7]

The doorbell signals a late arrival. As she rises to greet another guest, Maria might say with a charming Belgian accent, "Distraction can mean different things." Her husband browses in the World's Biggest Drugstore for amusement; she shops there for things they need.

Another guest weighs in. "What Miss Dunn said a moment ago

reminds me of the early Ch'an masters, before Buddhism migrated to Japan and became Zen. One monk rejected the idea of meditating for years and proposed the radical idea of Sudden Enlightenment."

Someone else offers his own koan: "If the present is a cow pie then the Buddha-mind lies under your foot."

Laughter ripples through the music room.

Television producer Roy Maypole, by this point in the evening, probably is fed up. All this arcane bantering adds up to zero potential for a TV program—he had expected a séance or table tapping or maybe age-regression hypnosis fakery, or some other trick he could deliciously exploit. He'd hoped to see the Huxleys' pet trance medium, Sophia Williams, who was said to communicate with her spirit guide in an eerie little voice. But none of this is happening tonight. Maypole is done with talk of cow crap and won't bother coming back.

As a general rule, guests did not bring young children to the Huxleys' on Tuesday nights. I stayed with my grandparents when my mother accompanied Dad to North Kings Road, which in her case wasn't often. But on one of those evenings after everyone had settled down in the music room and Maria was sitting next to my mother, the two women found a shared affinity over at least three things they had in common.

Both had an only child, although Matthew was a grown man and I was in grade school. Both supported their husbands by assisting behind the scenes and typing up their later drafts—Aldous typed his own first drafts, Howard printed his longhand—and organizing their daily lives, although certainly in the Eisenhower era such wifely support was taken for granted. Maria read aloud to Aldous and helped with his letters and proofs. She entertained their guests and, since with his flawed vision he could not drive, she drove him wherever he wanted to go.

My mother, Wanda, likewise, typed my father's books, bought supplies, helped with his photographic experiments, and cared for the liturgical garb he wore in the ceremonies at St. Francis-by-the-Sea in Laguna, starching and pressing his surplice and mending his robe. In short, Maria and Wanda were married to men with unusual interests who mainly lived inside their heads, sometimes on another plane.

The two women had another trait in common, and this was their intuition and their sensitivity to the feelings and needs of others. Aldous credited Maria with making him less detached and more human, and Maria recognized this same quality in Wanda.

The women built and maintained the bridge between the island of ideas and the needy isthmus of family and friends. While Maria and Wanda sat together that night in the music room—while Howard stood in front of the group and explained the latest findings of his hand study to the dozen or so guests—I am told that Maria leaned in to Wanda and whispered, "Howard could not have done this without you."

Maria may have been dwelling on this realization about herself and Aldous, especially since in late 1951 she had been diagnosed with breast cancer.[8] She had undergone a mastectomy and, following radiation, was told her cancer was in remission. Perhaps she intuited that this was not true, and even a fleeting thought raised the question of how her nearly blind husband would manage without her.

The radiation treatments caused sudden mental disturbances, she told friends, and took her to a frightening place. Aldous later described to Osmond what happened to his wife as a "feeling as if she were on the brink, or even over the brink, of madness."[9] She coped with her temporary bouts of otherworldly terror. She saw them as the hellish side of her spiritual reference point, the magic she had discovered in the expanse of the desert.

Aldous accepted that his wife's cancer was in remission, largely because she later masked its pain by saying she was suffering from the lower back pain of lumbago. He accepted, or preferred to accept, her explanation. Even in 1952, though she would not hear the worst news for another two years, she may have begun a quiet search for someone who might carry on as her successor.

One of those possible successors was present during another Tuesday night session on North Kings Road. This particular night a different group had gathered at the Huxleys' place, and at least three of those guests would have their hand photographs taken. My father's log book shows the names in attendance included Barbara Muhl and her husband, then Universal Pictures production chief Ed Muhl (numbers 397 and 398), and Maria Huxley's younger sister, Rose, who would also have her hand photographed that night.

For her part, and certainly more than her skeptical husband, Barbara Muhl is immersed in new therapies, healing and hypnosis, and what might be revealed about our psyches by interpreting the structure of the hand. Ed Muhl, whose head is focused on the tangible world of producing big-budget films, is mainly along for the ride. Tolerant of his wife's interest in the paranormal or the occult or the uncanny or whatever you want to call it, Ed is present on this night in 1952.

I picture that, when her name is called, Rose walks into the dining room and toward my father's camera, which is mounted in a black box about the size of an early Philco television set. It sits atop the Huxleys' heavy dining room table, the one also used for séances, and is positioned chest high with its lens facing down toward the tabletop. Howard finishes stirring the white, minty-smelling emulsion. Scents trigger memories, and possibly Rose's mind wanders. Now in her forties and a decade younger than Maria, Rose has survived two wars and two marriages. During the First World War she was evacuated from Belgium along with her mother and her sisters Suzanne, Maria, and Jeanne. Like Maria, she migrated to the States two decades later, before the Nazis began raining bombs down on Europe. She has a daughter, Olivia, and a son around my age named Sigfrid. They call him Siggy.

At this point, Howard takes Rose's hands and carefully coats her skin with the white chalky liquid. Satisfied that it has worked its way into the lines and pores, he places her hands in position below his camera lens and fiddles with the strobe lights. Rose waits patiently while he makes adjustments. Howard presses the shutter release and captures the image: Rose Nys Wessberg, hand photograph number 400. On this night, with the exception perhaps of Maria herself, no one in the music room imagines that in less than two years Maria Huxley will be gone. When that day arrives, some friends and family members will assume that sister Rose will become Aldous's support, comfort, reader, and helpmate, but they will be wrong.

To step back for a moment to the idea of support—life support, wife support, many kinds of support from emotional to therapeutic, secretarial to financial—if you are someone who lives inside your head and you have what you consider a great idea, then you might propose a project; you might submit a proposal to solicit funds from the honey pot controlled by nonprofit organizations called foundations.

One night, at a dinner party that took place in 1952 at the Huxleys' home, one of the guests was Robert Hutchins. He was head of the Ford Foundation, established by Edsel Ford in 1936 for "for scientific, educational and charitable purposes, all for the public welfare." Not yet a publicly traded entity, the Ford Motor Company in 1952 was still tied to its foundation and flush with revenue from the sale of autos purchased by returning GIs. As Huxley explained in a letter to his brother Julian, this meant that Ford was under pressure from the Internal Revenue Service "to treat accumulated revenue as current revenue, and so must get rid of 120 million dollars in a hurry." Moreover, Aldous added, "They have millions of dollars and no ideas."[10]

Aldous and Bob Hutchins soon found they held similar ideas about education. When Hutchins was president of the University of

Chicago he had supported the study of classic works as the foundation of a liberal arts education, and this led in 1952 to publication of the fifty-four-volume Great Books of the Western World. Huxley likewise believed education should provide the foundation for a lifetime of clear thinking, not prepare a student, especially in the postwar period, to become just a cog in the wheels of, say, aircraft manufacturing.

Huxley listened intently that night when conversation at the dinner party turned to talk of the Ford Foundation and its need to fund worthy projects. His mind ranged over many topics, thanks partly to Maria's daily and eclectic reading-aloud service. Later on he informally suggested a few ideas he thought worthy of funding, and Hutchins relayed them to the appropriate department, but by early 1953 several of those ideas had been passed over. Still, it was too early to tell whether one might eventually take root.

This informal exchange between Hutchins and Huxley took place around the time Maria read aloud in the evenings from the article in the *Hibbert Journal* by Smythies and Osmond about work they were doing with mescalin and schizophrenia. Aldous was intrigued by this new area of research. It seemed to connect the psychosis and hallucinations of *The Devils of Loudun* with the mysticism of *The Perennial Philosophy.* He was between major projects. This topic had the earmarks of a future book by Aldous Huxley.

PART II

ONE BRIGHT MAY MORNING

CHAPTER 7

TAKE A SIP OF AMAZEMENT

Huxley sent a personalized copy of *The Devils of Loudun* to Doctors Smythies and Osmond, care of Weyburn Hospital, Saskatchewan.

In Huxley's note accompanying the book he said he was very interested in these phenomena and had long wished to take mescalin, Smythies recalls. "He said if I planned to be in Los Angeles at any time, would I consider giving him some."

Smythies had no such plans, but his colleague Humphry Osmond would soon be attending a professional conference in Los Angeles.

Osmond personally thanked Huxley for sending the signed copy of *The Devils of Loudun*, but he didn't just stop with a thank you. He wrote back at length on March 31, 1953, adding his own opinion about how contemporary psychiatry barely scratched the surface. Osmond had recently read a critical review of *The Devils of Loudun* in some publication or other, and he told Huxley he strongly disagreed with the reviewer. Convinced that the writer had missed the point of Huxley's book, Osmond said he would write a review for a proper audience, the readers of what he called "our little psychiatric journal."

In early April the letter arrived in Huxley's mailbox, written in Osmond's quirky calligraphic style. Deciphering the handwriting, Maria Huxley read aloud to her husband: "I may be at the American Psychiatric Association [APA] meeting in Los Angeles this spring and would, if at all possible, like to call on you."

For Huxley this was very good news. Moreover, here was a kindred spirit willing to challenge the guard dogs of psychology and psychiatry, someone who believed, as Osmond wrote in the same letter, that "Aspects of human personality can be described in terms of Pavlovian dogs . . . other aspects can be described in Freudian terms, but when this has been done we are left with great continents of experience, with a stratosphere and a sub-oceanic region still untouched."[1]

Osmond, having a poetic streak, whetted Huxley's appetite by sending an account of his own recent mescalin experience, recalling how "a bird in the street, a sparrow, small and far away, might suddenly become the focus of one's attention, the most important thing in the world, the bird of the world, a key to the universe."[2] Huxley was eager to try mescalin and see that bird-of-the-world for himself. He wanted a personal experience of no-name pure consciousness, language being poorly suited to express the infinite even if that is what Aldous had tried to do in *The Perennial Philosophy*, when he linked samplings from mystical traditions over centuries and across many cultures.

Appreciating how eager Aldous was to try this experiment, Osmond cautioned him that inducing a psychosis is risky. He advised that should Huxley succeed in locating a source for mescalin, or a similar, synthetic, substance called lysergic acid diethylamide, he should not take it alone but with at least two companions. The experience, he said, would last eight to twelve hours. As an added precaution he listed substances that should be available in case a doctor had to terminate a session.

Moreover, if Huxley found a source and a physician willing to administer the drug, then he would ask a favor: Please record your impressions.

"John and I hope to interest a number of able people in this work and get them to record their experiences," Osmond explained. Their goal was to enlist luminaries in the fields of philosophy, literature, art, and science. Once collected, these impressions and opinions would not be ignored.

Osmond's news about attending the upcoming APA conference

in Los Angeles inclined Aldous to propose something unusual, but first he had to discuss it with his wife because the Huxleys rarely opened their home to overnight guests. When he suggested to Maria that they host Osmond, and after waffling a bit over worry that he might have a beard and be troublesome, Maria consented. In fact, later on Maria and Humphry Osmond would become close. At the time, though, Aldous wrote back to Osmond, "If you are coming alone to the meeting, we can provide a bed and bath."[3]

Thousands of miles away in Weyburn, the frozen Saskatchewan soil had barely begun to thaw. This was about the time Osmond opened his mail and found a letter from someone who, at that moment, was enjoying a sunny but windy 65-degree day in Los Angeles.

Dead brown grass and dirty patches of snow bordered the circular driveway leading to the formidable facade of Weyburn Mental Hospital. Inside the five-story brick administration building Osmond sat in an office with windows overlooking trees still bare, except for a pair of pines. Along with an invitation to be the Huxleys' houseguest, in the same letter Aldous had speculated about how mescalin might work on the brain. This was an idea borrowed from philosopher Henri Bergson—an idea Osmond and Huxley would chew over in letters and in person over the next months and years. Huxley's idea, as expressed in the April 10 letter, was this:

The normal human brain must act as a filter by selecting from among the "enormous possible world of consciousness," comprising both inner but mainly outer stimuli, so that it focuses on "biologically profitable channels." These profitable channels enable everyday decision making, such as finding provisions on a retailer's shelf, keeping to a train schedule, even closing the front door. In Huxley's view, this so-called normal brain can, at times, be overridden by events such as "disease, emotional shock, aesthetic experience, mystical enlightenment"—and now add to the list, ingesting mescalin.

Osmond accepted Huxley's invitation, writing back to express his thanks for the offer of a guest room and adding, half in jest, that he hoped not to be a nuisance. When he heard of the arrangement, John Smythies told his colleague the Oxford Professor H. H. Price (whom Smythies had mescalinized in 1952) the news that Osmond was scheduled to visit Huxley in May on a mescalin mission. Professor Price wrote back to Smythies, "It is very good news that Huxley is going to try mescal [mescalin]. He is just the man who ought to."[4]

Three more letters crossed the border prior to the opening day of the APA conference in LA. In one exchange, an early volley in a serious game that would continue for a decade, Osmond developed Huxley's point when he wrote, "In our culture, not only does the brain focus on the Sears Roebuck catalog and television screen, but our whole endeavor is to ensure that the Door in the Wall, which H. G. Wells described so wonderfully, is not only locked but its existence denied."[5] Apart from this phrase mentioned by Osmond, Huxley's literary repertoire included this, by poet and visionary William Blake: "If the doors of perception were cleansed everything would appear to man as it is, Infinite," a line from *The Marriage of Heaven and Hell.*

Huxley mentioned in another round of correspondence on April 19, 1953, that he had been unable to obtain mescalin through a doctor friend, who had tried to get it from Hoffmann–La Roche in Switzerland. He asked Osmond if he would bring some along. Osmond replied that he would be glad to bring the synthetic version of the cactus derivative, then ended his note with a light but telling touch: "I hope the narcotics people don't get on to me," he said, "but I expect that they will wink at anything for an APA meeting."[6] This suggests a mix of risk and caution on which Dr. Humphry Osmond seemed to thrive.

In many ways, Aldous had been preparing for a life-changing experience long before a Canadian visitor brought a vial of precious cargo in his suitcase. Huxley was conversant in historical references and scientific categories having to do with drugs. In *Brave New World*, he borrowed the idea of soma, the drink of the gods recorded in the ancient Vedic tradition of Indian philosophy, which no one knows what it was to this day. He was familiar with the work of German pharmacologist Louis Lewin and his five categories of psychoactive drugs and plants. Huxley wrote about this in his 1931 three-page essay called "A Short Treatise on Drugs." The same collection, *Music at Night and Other Essays*, included "Wanted, A New Pleasure," comparing speed to inebriants. He was familiar with Professor William James's 1902 book, *The Varieties of Religious Experience*, and with James's view of how taking an anesthetic like nitrous oxide could occasion a mystical state. He would come upon names of other peyote pioneers, names like Arthur Heffter and Havelock Ellis and Weir Mitchell.[7] Huxley was primed for a new adventure, one with the potential of providing new material, not knowing that his name would be added to the list of pioneers.

Conference week in May of 1953 began for Osmond with a flight along a southwest route on a silver fixed-wing, propeller-driven DC-3, which landed at the Los Angeles airport around two in the afternoon. Huxley had provided instructions for ground transportation, so after retrieving his suitcase Osmond rode the airline bus on this warm, smoggy day. The city of LA struck him as immense and sprawling, a jumble of skyscrapers along with miles and miles of what he described as "hermetically sealed air-conditioned palaces."

The airline bus took him to the Hollywood Roosevelt Hotel, located across from flamboyant Grauman's Chinese Theatre, with its sweeping temple entrance guarded by stone Fu dogs. Its Walk of Fame forecourt, scored with handprints and footprints of stars, dated back to the day in 1927 when the first two imprints were made by Norma Talmadge and Douglas Fairbanks. No mistake, this Londoner-turned-rural Canadian had arrived smack in the middle of Hollywood.

The Huxleys' home was located a couple of miles east of the hotel, from which a taxi took Osmond and his luggage to the Spanish-style home at 740 North Kings Road. When the Huxleys greeted him, Osmond was surprised at how friendly they were, and how pleasantly cool it felt to be inside their adobe-walled home on a warm spring day.

Luggage now stowed in the small guest room, he took a tour of the house and admired the Huxleys' art collection. Aldous had shown a proclivity to draw as a child and had often painted daily during the years in Sanary-sur-Mer. Throughout his adult life he had enjoyed visual arts as much as anyone, through the close-up aid of his everyday optics. Soon after his arrival, Osmond confirmed what he had already heard: Aldous had the use of only one eye and even that was impaired. Humphry Osmond wrote home on May 6 to his wife, Jane, saying, "He carries a magnifying glass for close work and a little spy glass for long distance."

The two men, one old enough to be the father of the other, may have had time before dinner to touch on a topic or two they had already begun exploring in their letters. One topic was the nature of hallucinations, another the mind-body problem. Osmond had noted in a letter how scientists rarely discuss mind-body-soul relationships, adding, "but I believe that they would become a very lively issue among a group of ex-mescalinized scientists."[8] The idea of scientists buzzing around mentally akimbo must have tickled Huxley, whose barbed pen had been largely responsible for his early reputation as a writer.

Osmond soon became aware of adjustments the couple had long since made to accommodate Huxley's near blindness. He observed his host's detached demeanor, his "well-bred cool," and in describing it to Jane said, "Aldous is completely vague and drifts slowly . . . like some elegant fish." In contrast, Maria scurried about and arranged her husband's affairs, which clearly he enjoyed. Even if she did not enjoy all the scurrying herself, Osmond could see that Maria felt appreciated. It was apparent how much Aldous needed her.

The next day the houseguest took a cab to downtown Los Angeles and the American Psychiatric Association conference at the recently completed Statler Hotel.

I had a chance to visit the scene of the 1953 APA conference half a century later, a few years before the hotel was replaced with a seventy-story skyscraper (the concrete pouring of which made the *Guinness Book of World Records*).

On the day I paid a visit, I reached downtown LA and navigated its one-way streets, then made a right and suddenly the facade came into view. Here was the aging beauty wearing her swoopy-hat porte cochere, the grand entryway found on the high end of midcentury modern architecture. I turned into the underground parking level of the Wilshire Grand, the former Los Angeles Statler Hotel.

It was easy to imagine arriving here for a vacation or convention in the early 1950s when, literally and figuratively, you had arrived. That's how it must have seemed in May of 1953 on the day when young Dr. Humphry Osmond walked into the American Psychiatric Association conference held in this futuristic hotel with its 1,275 guest rooms, the largest such structure built in the nation in more than two decades.

I had an appointment with the hotel's general manager, Marc Loge, who, fortunately for me, was a history buff. He escorted me into his office, eager to share his archives, and from a desk drawer produced photographs of the Los Angeles Statler as she appeared when brand new fifty years ago. I admired the turquoise arc of a roof jutting out over a line of vintage fishtailed cars disgorging passengers in the semicircular drive. Other photos showed bellhops carrying luggage, businessmen in suits and hats smoking cigarettes and escorting ladies clad in petticoat-inflated circle skirts or slim, midcalf pencil skirts, and all the women wore gloves.

Marc sorted through color photos and smiled as he slid another one across the desk. Seeing this juxtaposition of space age meets the South Pacific I said something like, "I feel like we should be holding tall drinks with paper umbrellas." I was looking at an image of the tropical pool enclosure studded with palm trees, imagining a 78-rpm recording of Sinatra's "You Do Something to Me" wafting through the pool area while I sunbathed in a lounge chair back in pre-ozone worry days when it seemed safe to do so. If lying on my back I would have seen, through cat's eye sunglasses, just a few modest skyscrapers looming on the new hotel's perimeter here in downtown LA.

After we talked for a while he wanted to show me something in the lobby. As the elevator descended, I detected a rising note of excitement in his voice and figured he didn't have many chances to talk with an enthused visitor about this historical stuff. We turned right outside the elevator and he guided me to a wall you'd miss if you didn't know where to look. It was filled with framed documents and photographs.

"Here we are," he said, with a voilà gesture. "The broadcast award banquets took place here in the 1950s." I was thinking how television was the new technology, perfect for a newly minted hotel. And sure enough, next to a photograph of the Statler Hotel's main banquet hall/ballroom I saw a commemorative parchment with the names of the 1953 Emmy winners for the year 1952: Lucille Ball, Best Comedienne; her counterpart, Jimmy Durante; Best Mystery, *Dragnet*; Best Children's Show, *Time for Beany*; Best Quiz Show, *What's My Line*.

I had told him on the telephone that I was researching a book about Huxley in the 1950s, but since most people associate Aldous with *Brave New World* I hardly expected the manager to connect a different set of dots. Pointing to another Emmy award year and the name Art Linkletter, he said, "I know you are writing about Huxley in the 50s and probably the mescalin story in *The Doors of Perception*." My face surely registered surprise as he went on. He asked if I remembered the alleged incident in 1969 when Art Linkletter's daughter

Diane jumped to her death from the sixth floor of her Westside apartment. I replied that certainly I'd heard about it, although I understood that no trace of LSD was found.

"It's an odd coincidence," he said. "That happens to be the same building where I now live."

Odd coincidences and strange dots, that was the shape this story was taking. Jogged by seeing the award for Lucille Ball for her starring role in *I Love Lucy*, I thought how I'd have to *'splain myself*, having become so wrapped up in Huxley and his experimental circle of friends that I had committed myself to what would turn out to be a decade-long writing project.

The mescalin session was scheduled for a day or so after Osmond's conference obligations were out of the way. When Osmond arrived for the conference, he had barely set foot inside the Statler when he was greeted with a flurry of congratulations. He was a newly elected member of a progressive subgroup called the Group for the Advancement of Psychiatry, or GAP. He later wrote to his wife that he was surprised because he hadn't realized what an honor this actually was.

On his second evening in LA, or perhaps on the third, there was another surprise. Osmond's visit apparently coincided with the Tuesday night experimental salon, or maybe there was some kind of an impromptu gathering, because when Osmond wrote Jane he said of the Huxleys, "They have a strange collection of friends. He always encourages others to have their say."

Chances are good that one of the strange friends that night was my father, who recalls being introduced to a visiting Canadian psychiatrist. Another guest that night, often a Tuesday regular and if so then the center of attention, was the previously mentioned psychic or trance medium named Sophia, whose vocalizations took the form of a tiny voice so strange that those hearing it, like Osmond and my father, said they would never forget it.

Communication like Sophia's, which if real must involve some unusual dimension of time and space, was at the philosophical heart of phenomena being studied at J. B. Rhine's parapsychology lab at Duke University. But Rhine's narrowly defined procedure of studying ESP, or *psi*, by testing subjects with Zener cards and statistically tracking results could not accommodate anecdotal or one-time uncanny incidents, no matter how dramatic or how many witnesses happened to observe it.

Rhine's procedure with Zener cards was not his first pass, because historically speaking his experiments dated back to his research working with trance mediums like Sophia. Rhine's Parapsychology Laboratory had been seeded with funds provided by trance medium Eileen Garrett, who in the '50s would become a friend of Osmond and Huxley, further expanding the experimental circle. Rhine's funding initially came from one of Garrett's séance clients, Mrs. Frances P. Bolton, an Ohio member of the US House of Representatives and one of America's richest women.[9] Bolton's son had died in a swimming accident, and, through what might later be called channeling, Garrett provided Bolton what John Smythies, who knew both of them, called "a great comfort."

Rhine's experiments since the 1930s ranged from working with subjects who simply, though still remarkably, accessed the thoughts of a person sitting in the next room, upward to experiments with individuals who claimed to predict the future, on to the farther extreme of those who like Eileen Garrett claimed to be able to communicate with the dead.[10] All three phenomena had become of interest to Huxley, especially since around 1950.

When Osmond returned the second day of the conference, he brought Aldous, who could be seen peering through his spyglass and enjoying, in his own unique way, an excursion to this rather stuffily self-important gathering. Amusing himself and Osmond during the general sessions, Huxley would genuflect every time Freud's name came up. His droll playfulness extended to taking a poke at prevailing political ideology, as this was the peak era of Senator Joseph McCarthy's anti-Red witch hunts. In the lobby, Huxley's distinctive voice carried over the conversational din, with remarks such as, "Humphry, how incredible it is in a Marxist country like this. . . ."[11] I am guessing most psychiatrists within earshot were shocked, unless they had read his novels and knew the bite of Huxley's humor.

After three days described by Osmond as "spent in a welter of psychiatrists" and nights spent at North Kings Road, both Huxleys and Osmond embarked on a shopping trip so Humphry could pick out gifts to take home to Jane and his daughter. The recently opened Ohrbach's Department Store sat on a prime corner at Wilshire and Fairfax in the Miracle Mile district, where many transplanted New Yorkers now lived. Ohrbach's carried a marked-down, sophisticated inventory displayed in an atmosphere of Manhattan-style bustle. Despite or because of his disparagement of the Sears Roebuck catalog, Aldous enjoyed perusing aisles crammed with goods, meandering in big stores bloated with selections like the Owl Rexall and Ohrbach's.

Osmond and Huxley strolled up and down the aisles, with Maria a few steps away. They passed accessory displays and clothing racks, as Aldous scanned the store with his spyglass and honed in on details with his magnifying glass. Maria, who drove the car and made sure Huxley didn't lose his way, helped Osmond pick out a mustard-gold coat for Jane, a striped coat for Jane's sister, and gifts for their first-born child, a little girl named Helen and nicknamed Tig (who would become the elder sister of a future baby girl nicknamed Fee).

The conference days passed, and the mescalin session loomed near. Osmond had his concerns. What if things went terribly wrong? What if he was the person to cause this famous author to go mad? Despite such worries, he wasn't the type to back down and Aldous was eager to give it a try. The mescalin experience might spark a new book. Mainly, though, Aldous wanted to experience the Ground of

Being first hand, or, as he would later write in *The Doors of Perception*, "It had always seemed possible that, through hypnosis . . . meditation, or else taking the appropriate drug, I might so change my ordinary mode of consciousness as to be able to know from the inside what the visionary, the medium, even the mystic were talking about."[12]

And so the following now-famous incident took place, according to Huxley and Osmond's account, on a bright day in May at the house on North Kings Road at around eleven o'clock in the morning:

> The Canadian psychiatrist shakes a capsule's worth of mescalin into a glass half filled with water and watches as the flakes dissolve. Humphry Osmond hands the glass to Aldous Huxley, who tips his head back slightly to drink, and they wait. Huxley has imagined closing his eyes, being swept away like a cosmic tourist to the realm of the visionary artists, poets, and mystics, the world of gems and soaring architecture, but this isn't what happens. No rubies or castle turrets, no encompassing radiance, at least not yet.
>
> Thirty minutes after Aldous has swallowed this liquid the everyday *I* begins to seep out into the room, or the room seeps into the *I* as he witnesses "a dance of golden lights"[13] appearing behind his own eyelids yet simultaneously floating in front of him, but the inner-or-outer I-and-Thou of it all is breaking down, and now a liquid-solid-gaseous red surface begins swirling, dissolving, re-emerging.
>
> Osmond, his guide, switches on the Dictaphone machine. It is important to record Huxley's verbal response to prompts, Osmond believes, and such prompts are needed because mescalin induces a solitary and mainly wordless state of mind.
>
> Things protrude in ultra-three-dimension, like the vase with three flowers. The psychiatrist sees that his subject is fixated and asks, "Is the sight pleasant to you?"
>
> "It just *is*."[14]

Huxley glances down at where his trousers form a lap, where they are lapped, lapped over, lapping not like water but *folds in fabric, folded into hills, into all the hills of the world.*

Books glow. Osmond sees Huxley staring at the bookshelf and asks, "What about time?"

"There seems to be plenty of it."[15]

"What about the room's interior?" the psychiatrist asks, referring to a wicker chair, typing table, and desk sitting nearby.

A still life—Oh, but the chair. Something about the legs! Later Huxley will write, "How miraculous their tubularity, how super-natural their polished smoothness!"[16] but that was Huxley writing as Huxley, and that expression came later.

Osmond leads Huxley through a French door and into the garden, where a white wooden lattice casts shadows onto chair cushions below. Aldous, seeing the patterns, pulls back, first in awe then in sheer but passing panic.

Around midway through the twelve-hour session he will be taken for a drive. The waiting car is parked at the curb.

Cartoon car. Bulbous top. It strikes him as so funny he can hardly stop laughing.

They motor up to the Hollywood Hills, where Osmond photographs Huxley standing alone and staring out over the canyon, before he remembers his subject's limited vision and how Aldous is unlikely to be enthralled by a sweeping view. Back they go into the car, cruise down the hill, and it is time for another change of venue.

Once back in the flats they head for the World's Biggest Drugstore, where they meander through the aisles, naturally gravitating to the book section. When Aldous pages through art books two pieces catch his eye. One is Sandro Botticelli's *Primavera*, its six female figures clad in diaphanous gowns.

Fabric folds, folds of fabric, the same as my trousers, the folds of the world.

The other is Vincent Van Gogh's *Bedroom in Arles*, with its straw-seated wooden chairs . . . *The same legs!* (Later he would write, "The chair Van Gogh had seen was obviously the same in essence as the chair I had seen.")[17]

Once back at North Kings Road, Osmond asks: "Do you know where madness lies?"

Aldous answers: "Yes."[18]

They round out the session with recorded music, and Osmond plays one of Huxley's favorite instrumentals, except this time it means nothing to him. Nothing, that is, until Osmond changes the recording to a choral piece, a late Renaissance madrigal by Gesualdo, and now Huxley pays attention. "These voices," Aldous will later write, "they're a kind of bridge back to the human world."[19]

I wish I could lay my hands on the Dictaphone recording that captured Huxley's answers, and sometimes I wonder if literary license filled the lines between the recorded words and those processed through the Huxleyan distill-and-synthesize process on the way to a book for publication. No matter, but I would expect his spontaneous replies to have been cryptic, because those who have ingested mescalin or its chemical cousin LSD know how hard it is, once in an altered state, to organize your thoughts. If you try to make notes then the ink trail of your pen is likely to devolve into curlicues.

Huxley isn't like the rest of us. With his command of language and reservoir of literary allusions, Aldous was primed to take this journey to an interior realm and make a travelogue out of what he'd found. After the session ended that day he would start the process of interpreting and elaborating, first in a report to Osmond, then an essay that would become a book published nine months later as *The Doors of Perception*, in which maybe one line from those seventy-nine pages sums it up: "I was seeing what Adam had seen on the morning of his creation. The miracle, moment by moment, of naked existence."[20]

Figure 2. Aldous Huxley in the Hollywood Hills after taking mescalin on May 6, 1953, leading him to write *The Doors of Perception*. Photo by Humphry Osmond. Courtesy of the family of Humphry Osmond.

CHAPTER 8

MESCALIN AND MARILYN MONROE

After his week in California, and once back home in Saskatchewan after fifteen hours in transit, Osmond followed up with a May 13 thank-you note to his recent hosts. Reflecting his new friendship with both of them and addressing it to both Aldous and Maria, Humphry said he hoped Aldous had suffered no ill effects from the mescalin and looked forward to "hearing your retrospective views upon the strange experience."

Maria graciously wrote back on her blue stationery around this same time in an undated letter. Yes, naturally Aldous had needed to recover a bit (from the mescalin) and they'd had to catch up (after having a houseguest for several days), but Osmond's visit was in no way a bother. Maria had only observed rather than participated in the mescalin session, but she intuitively understood what Aldous had undergone. Or, as she wrote to Humphry, "It confirmed in my case rather than brought a completely new experience," which is to say that it echoed what she already knew, what she simply called her "magic."

John Smythies had known both kinds of "magic"—the spontaneous kind, the white oval of light he'd seen in his London flat, and the chemical kind, which he'd ingested with caution and sparingly after a bad dissociative experience. He had stayed behind in Canada when Osmond came to LA, but while Huxley and Osmond were getting to know each other Smythies had not been forgotten. Aldous had inscribed a portrait photo with a personal message and sent it home with Osmond, who told Aldous that Smythies had been greatly pleased with the photograph.

Osmond also wrote in his post-trip letter that Smythies would be coming round to their cottage in Weyburn that night "to try our new ESP cards on Jane, who is usually a good scorer." A face-to-face meeting between Aldous and Smythies would have to wait a few years but, even remotely, Smythies would play a part in Huxley's experimental circle of friends.

After the mescalin experience, as they stayed in regular touch through correspondence and occasional meetings, Osmond would keep Huxley informed about other investigative drugs, including one many times more potent than mescalin—lysergic acid diethylamide, or LSD-25.

The discovery of lysergic acid is an oft-told tale: In 1938 a Swiss chemist named Albert Hofmann, employed by pharmaceutical firm Sandoz Ltd., was experimenting with derivatives of the fungi ergot while searching for a drug with an application in obstetrics. After two dozen tries he synthesized LSD-25 (which explains the number), but not until five years later would he become aware of its psychoactive properties, which happened after he accidentally absorbed some of the substance through his skin. This led to an intentional experiment and a famous bicycle ride home. The bike ride later came to be called history's first acid trip.[1]

Four more years passed before another self-experiment was conducted, this one by Werner A. Stoll at the psychiatric clinic of the University of Zurich.[2] The results led Sandoz to release the substance in the late 1940s as a research and experimental drug under the name Delysid. As stated on its prospectus, Delysid was indicated for "analytical psychotherapy and release of repressed material."

A second indication for Delysid was "experimental studies on the nature of psychosis."[3]

The blossoming correspondence between Huxley and Osmond began to mirror their search for the larger meaning of mescalin. It soon became clear that they were reaching beyond the psychological and psychiatric and diving into philosophical and spiritual implications, as if they sought the greatest possible number of connecting dots.

But such a quest needed serious funding. As Osmond saw it, the research had two facets. One was quantifiable, to be carried out in the lab at Weyburn. The other was a more qualitative approach involving anecdotal impressions collected as recordings and interviews with far-flung subjects who had undergone the mescalin experience.

They were cheered by knowing Huxley had an inside track with Robert Hutchins of the Ford Foundation, and at some point during Osmond's visit to LA this opportunity had been spoken about—the Ford Foundation was eager to plough funds into new and worthy projects. Energized by such a prospect, Osmond followed up in a letter by asking if Aldous had spoken yet with Hutchins about the mescalin expedition.

Then the psychiatrist went on to explain this in greater detail, including why he had recorded Huxley's mescalin session:

> We have a daring and entirely original project which we would like to submit to him [Hutchins] and also some other projects if his stuffy experts won't let him put Ford money on our best horse. Number One project is the one that I outlined to you in Los Angeles: a series of recorded mescalin interviews with 50–100 really intelligent people in various professions and occupations. Our object would be to explore the transformation of the "outer" world and the revelation of the "inner" world which occurs.[4]

Osmond continued his pitch, "It is evident that until this has been experienced it is largely meaningless, but once it has been experienced it is unforgettable." This research project would need funding to pay travel costs for a meeting of respected, mescalinized participants—the kind whose articulate opinions would be taken seriously, who could vouch for what Osmond called "the splendor and terror that exist just around the corner."

The project he proposed to Huxley needed a name to avoid repeating a description every time, and, thanks in part to Huxley's supreme wordsmithing, they called it Outsight. This stood for an idea they would return to again and again: Mescalin reveals an inner eyes-shut world, but the amazement lies in the transformation of what one sees "out there." The working brief, per Osmond, was "personal reflections on the experience of taking mescalin by 50 to 100 notable subjects in philosophy, literature and science."

Within a few weeks Aldous and Humphry had worked out the details of the prospectus, including conditions and limitations upon which they would insist: The interviews would not be conducted in a lab, contrary to standard scientific method. In order to accept Ford Foundation money, which they expected to be forthcoming (for surely, they thought, the Foundation would welcome the chance to fund such a uniquely high-concept project), it must be agreed that Aldous would sit on the Outsight advisory board. Huxley, jointly with Osmond and Smythies, would decide which individuals should be invited to participate. Osmond listed these points and more but in the end admitted to Huxley "it does look a little fantastic on paper."

It is spring of 2007, and my home file cabinet holds the precious digitized-then-printed copies of every letter and postcard Aldous Huxley ever wrote to Humphry Osmond, and Osmond to Huxley, between 1953 and 1963.

I take them out and place the stack in chronological order on my desk. I see strikeovers and the different shades of ink Aldous used for his signature and postscripts. I see notes he had written from trans-Atlantic liners and letters jotted on hotel stationery, adding up to a trove of a collection that helps me peer inside Osmond's ten-year friendship and a roller coaster of joint projects carried out with Huxley.

The letters give me insights into the man my father knew, though the relationship between my dad and Aldous was more like three years than ten. Still, there are similarities. There were mutual projects with promise—call them unfulfilled promise—that hit a wall in the 1950s because of cultural forces pushing against them. Both projects might have merit if revived today.

First there was the project Huxley, Osmond, and Smythies thought would become a cause célèbre, the one they called Outsight. Initiated in 1953, this was not the first cause Aldous had taken up. His commitment to the Peace Pledge Union had ended in failure a dozen years earlier. The Huxleys' home in the Mojave Desert, the former site of a utopian community called Llano Del Rio, hadn't worked out and they'd had to abandon it. Then there was Gerald Heard's Trabuco College, which Huxley helped plan, but after it had fallen deeply in debt it had been turned over to the Vedanta Society.

Then along came a cause worthy of idealistic, hint-of-utopia zeal.

The far-reaching purpose of Outsight was no less a goal than to advance human consciousness, and to this end draw attention to a chemically induced way of accessing some higher dimension. You could call it the Ground of Being, alluded to by mystics throughout the history of the world's religions, yet somehow it is related to the same creative force experienced by artists like Botticelli, Van Gogh, and Leonardo da Vinci.

But in no respect was Outsight intended to lead to unfettered access for all. Recalling the way Huxley and Osmond drafted the proposal to the Ford Foundation, Smythies said, "No idea of any possible or desirable recreational use of these hallucinogens crossed their innocent minds."[5]

Outsight more closely resembled an educational opportunity of the highest order, and for Aldous such a project linked theory and practice, ideas and action. Aldous gave his name and full support, without reservation, to a line of research he believed could one day accrue to the benefit of all. Another way of looking at Huxley's involvement is as a redemptive move, a chance to re-imagine tomorrow. Outsight might bring about a real world reenvisioning of his 1932 dystopian novel *Brave New World,* which portrays a soulless future in which a drug called soma is doled out to the population as a pacifier to preserve the status quo. Mescalin did the opposite. And along came Outsight, just when Aldous had started mulling over ideas for writing a utopian novel.

Months before, Bob Hutchins had asked Aldous to submit ideas, and he did so, though those suggestions went nowhere. This new and dynamic proposal should have worked like the proverbial charm, but Huxley and Osmond soon found out that support from the Ford Foundation was hampered by the politics of the organization.

To Osmond, this must have sounded like the retrograde stance of the doctors at London's St. George's Hospital, who had dismissed his early mescalin findings back in 1950. But he lived in a world where you worked around such hindrances and changed positions, which was how he'd paddled his boat around trouble before.

Aldous, however, was hardly sanguine. In his letters to Osmond he responded true to form, tarring the Ford Foundation as Mesozoic reptiles. He held out hope that Bob Hutchins might be able to support Outsight, even though he knew the Foundation was stuffily academic. A few months prior he had railed about this in a letter to his brother Julian: "I guess they must have started with injudicious appointments of dreary people . . . until now the whole affair is one vast fossil [*sic*] coprolite, in which the nominal heads of the foundation find themselves enclosed, like flies in amber."[6] In other words, Aldous considered the Ford Foundation a repository of dinosaur dung.

Osmond, not counting on any one funding source, considered that good publicity might still attract development funds, maybe cause the Ford Foundation a change of mind, and he was willing to go to considerable lengths for the sake of this possibility. Mid-June of 1953 found him driving across the prairies of Saskatchewan on a three-day road trip to Winnipeg, where he supervised an LSD session requested by science journalist Sidney Katz, who wrote for the popular Canadian magazine *Macleans*. But Katz was unprepared for the nauseating physical side effects that often accompany such a session.[7] Smythies himself had even undergone a bad experience with mescalin, as he mentioned in a note after our first interview, calling it an unpleasant "instance of derealization." Aldous, who was corresponding with Smythies at the time, had encouraged him to give it another go, and he was right because Smythies's second time with the drug was benign.[8] But for Katz, the writer at *Macleans*, the outcome was a not-very-positive article that ran in October 1953 called "My Twelve Hours as a Madman."

Around the same time, although the timetable is a bit unclear and I'll take this point up again later, Osmond and his colleague Dr. Abram Hoffer are said to have begun experimenting with LSD as therapy, and patients admitted for chronic alcoholism comprised a large percentage of the population at Weyburn Mental Hospital.[9]

Aldous had been spared a bad experience in May, and that summer he wrote up a report of his positive mescalin experience and sent it to Osmond. When further expanded, it would be published in book form the following spring as *The Doors of Perception*.

Huxley's complementary project was his adjunct involvement in Outsight. After discovering that Bob Hutchins did not have veto power over naysayers at the Ford Foundation he and Osmond set about fortifying their case. Aldous asked Osmond to provide further details about how Outsight would be administered, figuring they needed a new approach for Ford and might propose the project to the Rockefeller Foundation.

Osmond laid it out in a rough draft for Aldous. The plan involved a flexible and simple method similar to their recent session at the house on North Kings Road (anticipating what would later be called setting). "Not too formal and formidable seems to me what we should need," Osmond said. He was looking for a few more names to add breadth to the proposed roster of gifted people he would like to "mescalinize." For his part, with Smythies help, he was aiming for Oxbridge philosophers A. J. Ayer and Gilbert Ryle, and hoping for more notables in arts and letters. "I would like to try Graham Greene," he wrote to Aldous. "Do you know him by chance?"

In theory, once they had built up a group of gifted people willing to participate, and once they had individually administered the drug so each had direct knowledge of what might be called a transcendent experience, then the next step would be to gather them all in one location—a kind of mescalinized think tank.[10] With so many analytical and creative minds in one place, Osmond and Huxley believed they could communicate the importance of this discovery to the larger world. Aldous was a single admired voice soon to go public, with almost missionary zeal, in *The Doors of Perception*. Outsight would be the voice of many.

Filling in specifics, as Aldous had requested, Osmond estimated Outsight would cost perhaps $35,000–$40,000 for the first two years. It was affordable because the principal researchers would be able to use existing facilities in Weyburn, also since they were employed by the province of Saskatchewan rather than in private practice it would amount to half the cost should this be executed elsewhere. There would be travel expenses, of course.

Aldous, who would help polish the final draft, hadn't given up on the Ford Foundation, and they thought Hutchins might want to take the stuff. "If Hutchins would like to take mescalin, I'm sure that John S. and/or I could come down, somehow," said Osmond.[11]

But one of their partners in this enterprise had decided to break away. John Smythies announced that he would leave Weyburn to accept a staff position in the Department of Neurological Research at the University of British Columbia, where he would continue his work in basic neuroscience. Osmond was upset at first about what he saw as a defection—John leaving Weyburn at a critical time for Outsight! This rubbed off on Aldous, who was not happy about it either. But the resentment soon abated, and collaboration with Smythies would continue for years. Moreover, John's Vancouver base would lead to a contact that before long would become important, for better and for worse, in Huxley's experimental circle of friends: Captain Al Hubbard.

Before Smythies left his position at Weyburn Hospital, he, along with Osmond and their colleague biochemist Dr. Abe Hoffer, recast the Outsight proposal as an eight-point outline and sent it to Huxley for review. It arrived with Osmond's note asking Aldous to modify or "suppress anything which you think is unsuitable."[12] To further fine-tune this prior to submitting it again to the Ford Foundation, Huxley showed Gerald Heard the draft and asked for his input, too, just as Heard had asked Huxley for help with the Trabuco College prospectus a dozen years earlier.

Word spread about their plan. Some of Huxley's friends with experimental leanings offered to create a fund to bring Osmond to Los Angeles for a group session, anticipating that this session might include the Ford Foundation's Bob Hutchins. Spirits were rising. The mescalin think tank idea was looking more viable, and another name was added to the roster when around this same time John Smythies received a note of encouragement from the notable philosopher

H. H. Price, who wrote, "How interested I am in the project you are submitting to the Ford Foundation and I should be very glad indeed to be an adviser along with Aldous Huxley."[13]

Two kinds of documents shuttled back and forth across the United States-Canadian border during summer of 1953—the Outsight proposal and the manuscript, the latter on its way to becoming a book. As the publication process churned on, Osmond and Smythies sent suggestions to Aldous for revising the draft of *The Doors of Perception*. After incorporating the changes with which he agreed, Aldous mailed the final manuscript to his editor, Cass Canfield, at Harper & Row, and the book was scheduled for publication in February 1954. Canfield arranged for the essay to appear in serial form, a typical subsidiary rights deal in which a book first appears in a magazine to build interest in the forthcoming book.

Huxley had written for Condé Nast's *House and Garden* magazine early in his career, but this magazine was of a different stripe. Huxley playfully informed Osmond that the mescalin essay "is to appear seri-ally—of all places—in *Esquire*—which is at present engaged in serving God and Mammon, Petty Girls and moderately serious literature, with what I understand to be remarkable success. The P.G.s pay for the S. Lit. and both ends of the central nervous system, the cerebral and the sacral, receive their appropriate stimulation—to the satisfac-tion of everyone concerned."[14]

Osmond, who certainly chortled when he read this, sent a return volley: "I am greatly entertained and thrilled at your news," he wrote Aldous. "Modern journalism is a real hotch-potch—sex-appeal and schizophrenia, mescalin and Marilyn Monroe. Please let me know when the *Esquire* piece is coming out so that I can get some pin up girls."[15]

The choice of publication made sense, and the sub-rights income would be substantial. *Esquire* was a successful magazine, appealing to many returning servicemen who took advantage of the GI Bill to obtain a college education and, as the saying goes, to better them-selves. The mix of sex appeal and articles by the likes of Aldous Huxley smartly fit the bill.

Publicity from *Esquire* would be important for sales of the new book, too, because *The Doors of Perception* was far from what many Huxley readers expected. Aldous had a following for his eclectic essays, but his larger audience wanted a wry, sharp-tipped novel like his 1939 *After Many a Summer Dies the Swan*, or the mix of poignant reflection and cynicism found in his 1944 novel *Time Must Have a Stop*. In a way, Huxley and his readers were evolving together, but *Doors* would be the most extreme departure for him thus far.

Those witty exchanges mailed between Los Angeles and Weyburn proved to be premature glee, for it turned out that the magazine's cover date conflicted with the book's February release, and soon the magazine deal was pronounced dead.[16] Lost income was nothing to take lightly, and Osmond offered his condolences. Disappointed for his friend, he also had a stake in publicity surrounding Huxley's mescalin essay, for Osmond's name and his work with schizophrenia would be acknowledged in a prominent footnote in *Doors*. He wanted success for all three—Huxley's book, the Outsight project that he called their best pony, and the Saskatchewan Schizophrenic Research Group in Weyburn, his *other* pony. He wanted all three to thrive, and he wanted to bask in the glow of Huxley's byline.

As consolation, Osmond offered to write a preview piece in a dif-ferent magazine, although one with a far smaller circulation than *Esquire*. This other magazine, called *Tomorrow*, also offered editorial content popular in the 1950s, with a mix of health advice and ghost stories, science and science fiction.

Publisher Eileen Garrett, who has been mentioned before in connection with J. B. Rhine and his Parapsychology Laboratory at Duke, was someone who had claimed to experience visions since her early childhood growing up in Ireland. According to some, she also was one of the most respected and closely studied mediums of the twentieth century. Her reputation stemmed from one allegedly stunning incident of clairvoyance that occurred a few years before the onset of the Second World War.

The occasion was an evening séance around the time of the 1930 crash of the experimental British airship R-101, an accident that killed all on board. As the reportedly well-documented account goes, Garrett was sitting with observers in London during a trance session when the dead captain of the airship, Flight Lieutenant H. Carmichael, began speaking through her in what is called a "drop in," or unsolicited contact. A shorthand record was made at the time, and afterward the technical details, until then unknown by the public, were said to match information uncovered by the British Air Ministry during its investigation of the crash. The uncanny coincidence was widely reported in the British press.

Then, in 1931, Garrett was invited to New York by the American Society for Psychical Research to participate in a series of experiments administered by Hereward Carrington (hand number 246). While in the States, she participated in studies with J. B. Rhine at Duke University, before moving to New York where subsequently she quietly launched *Tomorrow* magazine. Initially it was a publication focused on literature and public affairs, but that was to change because high and low cultural forces were converging.

The pulp fiction periodicals of the Great Depression years lasted until after World War II, while at the same time literary periodicals continued to serve elite college-educated readers. The GI bill and returning soldiers heading to college, combined with innovation in science and technology, had an influence on popular reading material. In the publishing world of the 1950s, the last phase of the pulp fiction era produced strange offspring: True crime morphed into the paranormal, horror cross-fertilized with both science and science fiction. The pulp magazine *Weird Tales* in 1950 featured a cover story on L. Ron Hubbard's Dianetics. The relaunch of Garret's magazine in 1952 reflected this era of postwar GIs and their varied interests.

A decade after starting her literary journal, with underwriting from her wealthy patron, member of Congress Frances Bolton, Garrett founded the Parapsychology Foundation and relaunched the magazine as *Tomorrow: The International Journal of Parapsychology and Occult Sciences*.[17]

From this time forward, Garrett would gradually become involved with Huxley's experimental circle of friends. Members of this group who would write for the magazine during the early 1950s included Huxley, Gerald Heard, Christopher Isherwood, and hypnotist Leslie LeCron. Isherwood contributed the first book review of *Martian Chronicles*, a science fiction story collection by a new young writer named Ray Bradbury. Humphry Osmond's review of *The Doors of Perception* would appear in *Tomorrow* in early 1954, along with an article by John Smythies writing in the same issue about mescalin.

Osmond explained in the article he wrote for *Tomorrow* how for the past two years the Saskatchewan Schizophrenia Research Group had been experimenting with the alkaloid mescalin, found in the cactus peyote. "Our studies on the psychological effects of mescalin suggested to us that the visionary, the dreamer, the artist and the medium, the mystic, the schizophrenic and the mescalin-taker have experiences which have much in common," he wrote in the review, calling this "a miraculous subject." Ending his preview on a note of advocacy he wrote, "Huxley ends by challenging our chemists to produce a safe, cheap, and easily synthesized hallucinogen, and our religious and philosophical leaders to put this to a proper social use."[18]

CHAPTER 9

SLAMMING THE DOORS

Osmond's advanced copy of *The Doors of Perception* arrived in the Weyburn Hospital mailroom in January 1954, a month before the official publication date. He opened the brown postal wrapping. Reading what Aldous had written to him personally, crediting him with co-parenthood, Osmond wrote back with gratitude and humility, though insisted that his role was more like that of a midwife.

There had still been no luck with the Ford Foundation, of which Huxley had written to Osmond, "I have given up all hope of these bloody Ford people. They are obviously dedicated whole-heartedly to doing nothing that might look in any way novel or unorthodox."[1] In a letter to his brother Julian on the subject of the Ford Foundation, written on same date, January 25, but a year before, Aldous had called the foundation's decision-makers—or gatekeepers—"too stuffily academic," a phrase that would resurface under curious circumstances half a dozen years later.

Although the intended serial in *Esquire* was cancelled, Huxley's name was highly visible nationally in the month before release of his new book. The January 11 issue of *Life* magazine ran an article called "A Case for ESP, PK, and PSI," with a black-and-white photo of Aldous looking contemplative. The other photos showed J. B. Rhine and a young female subject sitting at desks and facing each other but separated by a wall. The caption read, "To test ESP, Dr. Rhine turns cards in one room as a girl in another records her guesses on their order. She flashes light to signal for turn." In the article, Huxley talked about Rhine's work and that of the Society for Psychical Research, and he made the case for ESP in part by drawing on documented accounts he called "on the borderline between the normal and the paranormal."

His examples included prescient dreams, in some cases simultaneous with the event and in others of someone having foreknowledge of an incident, such as someone dying in an accident. In one incident a mother remotely sensed the time of her soldier son's death, and in another a woman dreamed that someone's runaway pig ended up in her dining room. The next day it happened. One of Huxley's points in the article was that people seem more inclined to accept ESP, or uncanny mind-to-mind knowledge, than psychokinesis, or the direct action of mind on matter (including the body), a category that includes psychosomatic illness as well as remote healing.

When *The Doors of Perception* reached the marketplace the following month, in February of 1954, and with Huxley's name on the cover, the public snapped up copies. *Doors* arrived in bookstores, department stores, and libraries. Osmond later heard, and passed along to Aldous, that the noted Zen Buddhist scholar Dr. Daisetz T. Suzuki, who taught at New York's Columbia University, was insisting that his students read it.[2]

"Good news about the sales of the book," Aldous wrote to Harold Raymond, publisher at Chatto & Windus, his publisher in Britain, "excellent, I should say, for an essay."[3]

But *Doors* garnered mixed critical reviews and a few outright pans, some scholarly and some personal. One prickly review by medical writer Berton Roueché appeared in the *New Yorker:*

> It would be too much to say that the metaphysical convulsions that seized Mr. Huxley under mescalin excessively try our patience. He

is usually, in this as in all of his more supersonic flights, well worth attending. One might wish, however, considering the unusual nature of his spree, that he had given us a somewhat briefer review of his current intellectual preoccupation and a somewhat fuller account of exactly what happened, objectively and mechanistically, on that recent bright May morning.[4]

The Doors of Perception created divisions within Huxley's own camp, as well. A different kind of objection came from Cambridge philosopher C. D. Broad, a prospective Outsight mescalin think tank member who had been recruited by Smythies. Broad wrote to Smythies, "I thought that Huxley's book on mescal [mescalin] was somewhat irresponsible, but I suppose it cannot do much harm, as it is difficult for the public to get hold of the drug."[5] The observation was true at the time. In one letter to Osmond, Huxley had mentioned that it was nearly impossible to obtain mescalin or LSD in Los Angeles, but those were the early years; by the end of the decade both Huxley's and Broad's remarks would seem quaint.

Other respected names also took issue with Huxley's latest book. What bothered some was an implied assurance that enlightenment could be chemically induced. One such critic was Swami Prabhavananda of the Vedanta Society in Hollywood, where Aldous had guest-lectured and also had contributed many articles to the periodical *Vedanta and the West*. Reportedly, in Swami's view, Huxley's book was a guidebook not to a higher plane but a seductive one, a detour to a place of false side effects. Known in the Vedanta tradition as *siddhi*, the powers acquired through practices such as meditation are often dismissed as distractions.

Another objection came from across the Atlantic, this one from a comparative religion scholar, R. C. Zaehner. He was already on record as opposed to the idea of universalism—the belief in one common truth—that infused Huxley's book *The Perennial Philosophy*.

The Oxford professor had tried mescalin but afterward wrote it off as an experience of lower-order phenomena. Smythies, who had supervised Zaehner's mescalin session, said he thought the professor had not really achieved the full measure of the experience.[6]

Of course, Huxley did not claim that his day spent under the spell of mescalin brought on a state of enlightenment, though some might interpret it that way; he didn't say he had seen the face of God nor did he pull into the parking lot of paradise. But I can understand why Swami Prabhavananda, R. C. Zaehner, Huxley's friend Jiddu Krishnamurti, and others with reputations in the field of religion might reject taking mescalin as a quick route to transcendent wisdom, and why they would resent Huxley, with his bully pulpit of readers, for suggesting this was possible. In Huxley's next book he would develop the idea of visionary experience as a gateway— although not guaranteed—to a full mystical experience.

One reason for the leap to a conclusion that mescalin is a spiritual panacea is the language Huxley uses to describe his mescalin experience in *Doors*, as if piped in from *The Perennial Philosophy*. In his earlier book, Aldous was a curator of quotations and metaphorical images alluding to mystical experience found in the vast archive of the world's sacred scriptures.

If there is a problem of communication in *Doors*, when Huxley tries to convey the nature of his experience, it is parallel to the problem of describing mystical experience in any religious text—the limitation of words. That is the trouble with a trip to the Ground of Being—how do you tell others what you saw, except perhaps poetically, or is seeing even the right sense to rely on for such a report?

Huxley tried to do this by putting a gloss on the raw and immediate experience of a flower and a chair and a grid of shadows, but part of *Doors* is about history and theory, including educational theory, which comes back to the *New York Times* book critic's objection, when Berton Roueché wanted to know more about "exactly what happened."

John Smythies, for a different reason, wondered why Huxley had veered away from language close to the direct experience itself. It was one thing to describe the phenomena, which Huxley could do so poetically well, another to spend so many pages explaining it in religious terms and mainly those of Hinduism and Buddhism. Smythies wrote to Aldous suggesting that this was a mistake. "Surely, as an organization, our Group must present a front to the world as having no particular axe to grind?"[7] Apparently Aldous pushed back in a reply. Then Smythies backpedaled with, "When I said that you had made mistakes in *Doors of Perception*, I did not mean from your own point of view—the book gives an accurate analytical account of the situation—but I thought that they might be tactical errors from the point of view of our long-term aims."

Those aims, of course, included appealing to a large and secular Foundation so that its board might agree to fund the Outsight project.

I suppose if Aldous had written *Doors* as a magazine essay he might have stayed closer to the shore of descriptive reporting. Instead, he worked all volumes of his encyclopedic mind, and turned an essay into a seventy-nine-page hardcover book. I wonder if the length or content of the essay was ever an issue with Cass Canfield of Harper & Brothers, who, although chairman of the firm, was often in direct editorial contact with Huxley about the content of his books.[8] The record doesn't tell us. But the book caught the public's attention and by late spring of 1954, the score for *The Doors of Perception* was public opinion 2, critics of various stripes and persuasions, 1.

Critical judgment was on my mind on February 26, 2007, the morning I had my second appointment with Dr. John Smythies at the Center for Brain and Cognition at the University of California, San Diego.

Manila files lay strewn across my desk that day, their beige tabs like tongues scorning my approach to research. My home office looked like the aftermath of an earthquake. I was barely managing the bounty of original documents sent to me from Smythies in San Diego, from Birmingham, Alabama (where Smythies's papers are preserved at the University of Alabama), and from Wisconsin, Fee Osmond's home, not to mention notes from the material I had uncovered at UCLA so far and three shelves of used books and magazines I had collected pertaining to Huxley. I knew I could not incorporate all this source material into the flow of the story, nor do justice to the accomplishments of Smythies, Osmond, and others in Huxley's circle, much less draw a proper picture of a polymath like Aldous. I could only follow the arc of this story and tell it as best I could.

The materials inside my shoulder bag as I left to see John Smythies that day included a copy of the January 29, 2007, issue of *Time*, which had devoted nearly fifty pages to their Mind & Body Special Issue, subtitled "The Brain: A User's Guide." Most articles in this special issue supported the body-centric monist viewpoint, which holds that the mind is an imaginary construct and "I" is a brain driven by the body's chemistry. Not surprisingly, the issue was packed with pharmaceutical ads.

But the last article in the special issue, in a way reserving this page for the last word, stood in contrast. It was preceded by a quote from evolutionary biologist J. B. S. Haldane, who coincidentally had been a family friend of Huxley's parents. The quote in this issue from Haldane reads, "Now, my own suspicion is that the universe is not only queerer than we suppose, it is queerer than we *can* suppose."

The closing article in the Mind & Body Special Issue was contributed by a doctor named Scott Haig, an assistant professor of orthopedic surgery at Columbia University College of Physicians and Surgeons, who described in a first-person account the following incident: Haig had attended a patient in the advanced stage of a lung cancer that had metastasized to his brain. The patient, who was named David, had stopped speaking and moving, and the brain scan showed, as Dr. Haig described it in *Time*, "barely any brain left." The day after seeing Haig, David died. An attending senior nurse and David's wife, also a nurse, had been with David at the end.

The two women separately told Dr. Haig how, remarkable though it may seem, the patient had woken up and, for about five minutes, spoke to the family and touched them and smiled, then lost consciousness again and passed within the hour. Judging only by the brain scan, this feat seemed impossible.

Or it seems impossible in the monist viewpoint, which says that behind human mental processes the brain is all you have. In the dualist view, mind is not synonymous with the physical brain, the latter being constrained by time, as we know it, as well as by space or location. If mind is not identical with the brain then in theory it might be free of space-time limitations; it might be capable of action apart from physical deterioration evident in a brain scan, and this might account for the phenomena witnessed in David's final minutes. It might account for telepathic communication.

Before setting out to see Professor Smythies that day, I dropped a small recorder in my bag and hitched a camera strap over my shoulder. After about an hour of densely darting traffic I reached UC San Diego and followed a path to the pair of buildings housing the psychology department. When I reached his office, Smythies stood to greet me. When I'd last seen him he had been wearing a leg brace from knee surgery, and at that time I hadn't noticed that he was quite tall and physically somewhat like Aldous.

On this day he looked like an informally dressed Oxford don in his tan slacks and a red pinstripe shirt worn under a gray V-neck sweater, his blue eyes bright and lively yet red-rimmed, as though youth and age had come to an agreement. I asked about his knee. "Are you taking physical therapy?"

"Swimming," he replied, and I told him I also lap swim, a far less impressive endeavor for me than for this man entering his ninth decade.

At one point in the conversation I mentioned that my husband scorned parapsychology and subscribed to *Skeptical Inquirer*. "It seems to me," I said, "that such publications dismiss research into psi phenomena as either fraudulent or studies that don't meet the scientific standard." Smythies said that for several decades, reputable institutions, including the University of Edinburgh and the University of Virginia, have maintained departments devoted to parapsychology. Whether the results of studies provide evidence of ESP is an ongoing debate, with some studies showing a small but statistically significant effect size.[9]

How do the ideas of monism and dualism fit into a discussion of parapsychology? You could call them an underpinning of the discussion. One of many topics in philosophy is theory of mind, or the mind-body problem, which is not so much laboratory-testable as a way to talk about underlying ideas.

John Smythies takes the dualist position that the mind is not identical with the brain. This is the philosophical position he has held since around 1950, after the life-changing experiences of seeing the white oval of light in his room, after the convincing test with the clairvoyant hospital patient who knew what was inside a sealed letter and by whom and where it had been written. Smythies not only wanted to understand how the first was physically possible, which had led him to study hallucinations, but also why clairvoyance upends the accepted notions of space and time.

Today, the study of physics looks at many possible dimensions. Smythies has long grappled with the idea of mind extended in a spatial dimension, and I have seen this discussed in letters between he and Aldous. Which brings me back to Huxley and how that day in John's office I had to steer the conversation back to matters related to Huxley's life. At one point when I made this shift, Smythies looked up at me and said, emphatically, "Surely you must write about Le Piol."

CHAPTER 10

THE FRENCH CONNECTION

After the second interview with John Smythies I was relieved to have new leads in pursuit of half-century-old history. An hour later, back at my desk, I saw that John had e-mailed me a black-and-white photograph showing a dozen or so men and women seated in small groups on an airy veranda. Some were leaning back and apparently engaged in light conversation, others leaning forward with signs of intense concentration.

When I look at that photograph these days I imagine cicadas providing the soundtrack of a particular time and place: the two-part 1954 Symposium on Philosophy and Parapsychology that took place in the south of France at a private hotel called Le Piol. Aldous Huxley's first visit came in late April and early May of that year, two months after publication of *The Doors of Perception*.

Le Piol was the property of the internationally known trance medium Eileen Garrett and served both as her second home and as the setting of the Parapsychology Foundation's annual conference. Every year an elite group of philosophers, psychologists, and physicians converged on St. Paul-de-Vence, a Provençal village high on a rocky outcrop encircled by medieval ramparts. Overlooking the Mediterranean and the Cap d'Antibes, St. Paul is known for its unusual light.

Le Piol was the Huxleys' destination that spring, but first came Manhattan. The Huxleys had booked their trans-Atlantic passage on the RMS *Queen Elizabeth*, where the port of debarkation meant a chance for Huxley and Humphry Osmond to see each other in New York. Their frequent letters to each other were brimming with ideas, but replies involved postal delivery delays—though it was not as if their ideas grew cold. In what would be their second face-to-face meeting, they could explore mutual interests—ESP and hypnosis, the state induced by mescalin, and the thin line between mysticism and madness.

It was easier for Osmond to find excuses to travel from Saskatche-wan, Canada, to New York than to Los Angeles, and it happened that the annual psychiatric conference would take place in nearby New Jersey. Osmond added on days by lining up appointments for the purchase of much-needed cleaning equipment to improve what he had inherited as the director of Weyburn Hospital, where the most severely disturbed patients lived in a constant state of barely con-tained filth. Apart from those business meetings he would have free time to spend in the city with Aldous. It would provide a chance for Jane Osmond to meet Maria Huxley and the famous author who had become such an important part of her husband's life.

Aldous had reserved rooms for both couples at the Warwick, Huxley's favorite New York hotel. Located at the corner of Sixth Avenue and Fifty-Fourth Street, it offered a convenient location and the afterglow of a scandalous past.

William Randolph Hearst had commissioned the Warwick in 1926 for his ladylove Marion Davies. Meanwhile, Hearst's estranged wife, Millicent, resided in resentful opulence elsewhere, at one or another of the Hearst's far-flung estates. The Warwick Hotel's architect Emery Roth had created a Beaux Art–style beauty typical of the era's urban design on land across from the Ziegfeld Theater, where in their luxu-rious suite Hearst and Davies liked to entertain their Hollywood friends. That same year, just three blocks away, Hearst built the International Magazine Building to house twelve of his publications, among them magazines occasionally carrying articles with Aldous Huxley's byline.

I imagine Huxley appreciated the other scandalous tale associated with the Warwick. This concerned a suite of decorative murals adorning three main walls inside the hotel restaurant, each one celebrating historical events associated with Sir Walter Raleigh. One portrayed the apocryphal tale of Raleigh throwing his cloak atop a puddle so Queen Elizabeth I would not soil her royal pumps. The richly painted tableaux, generous in use of red pigment, included one scene showing the Queen showering Raleigh with gifts while New World natives stoically looked on.

But when the murals were almost done, as the story goes, Hearst and muralist Dean Cornwell argued over the amount of his fee. Hearst wouldn't negotiate, so the artist committed poison-brush retaliation by painting bare buttocks on a native and, worse, adding a male figure copiously urinating on the Queen. That last flourish so enraged Hearst that he ordered the north wall mural painted over.

Forty years later, the spite-inspired section of Cornwell's original mural would be partly restored, though the urinating soldier was now camouflaged behind a voluminous cloak. The Native American's buttocks are on view, though, and you can consume an expensive breakfast, as I did recently, in the restaurant now called Murals on Fifty-Fourth. The Warwick's wrought iron and brass entryway has welcomed guests for almost a century and commands prime real estate near the theater district, as William Randolph Hearst had always intended. Oddly alluding to its origin, it is billed as "A blend of grandeur and intimacy."

Though the scandalous mural had long since been painted over when the Huxleys and Osmonds stayed at the Warwick in 1954, other sections were visible. Aldous might have pointed out a small mural set apart from the others. This one is still found over an archway in the southwest corner, an image that does not jump out at you like the large tableaux with their celebration of Imperial might. This little mural portrays a brooding medieval alchemist bent over his magic potions, a more Huxleyan kind of dominion.

That April of 1954 the hotel provided a base for Aldous to show Humphry and Jane around Manhattan. He whisked the Osmonds off on a tour reflecting Huxley's many faces: Aldous the curious author, the avid art connoisseur—he was already working on an essay about visionary art—and the mock-the-commercial-world Aldous. They strolled past shops on Madison Avenue, where Jane and Maria enjoyed window displays, while Aldous and Humphry likely exchanged wry remarks about fleeting fashions. With his spyglass and magnifiers at the ready, Huxley escorted the Osmonds to the Museum of Modern Art and across town to the Metropolitan, where he pointed out paintings by artists in whose work a glimpse of some other dimension shone through: Hieronymus Bosch, French symbolist painter Gustave Moreau, and the man perhaps most often associated with the phrase visionary art—poet and illustrator William Blake.

Since both couples had other commitments as well, after spending a few days together in the city they went separate ways. Osmond resumed his hospital director's role and met with the supplier of a new kind of bed, this type being easier to clean, and placed orders for the hospital. Afterward he and Jane continued on to the psychiatric conference in New Jersey.

Aldous had a few social engagements in the city, among them meetings with people presumably powerful enough to influence the Ford or Rockefeller or other foundations, and put in a good word for the mescalin project. Seeds having been planted, a few days later the Huxleys boarded the RMS *Queen Elizabeth* bound for Cherbourg.

Whether on land or at sea, Aldous typically wrote several letters a day. One written that week to Osmond on *QE* stationery informed Osmond that they were somewhere in the mid-Atlantic where "all is well here, sea like a mill pond . . . steady eating and sleeping, with a bit of work in the rare intervals." Aldous added a postscript: "I hope something has come, or will come, of my activities among the Rich & the Foundations."[1] Aldous also wrote to British playwright Clifford

Bax (who had an interest in esoteric matters, might be a potential supporter, or could know others who might wish to be supporters), explaining the project spearheaded by Osmond and adding, "Needless to say, several Foundations have turned him down; but we have not yet lost hope. Someone may see the importance of the project and put up funds—trivially small by comparison with what is spent in other fields—for exploring those Other Worlds which we all carry about with us."[2]

The Atlantic crossing took five days, followed by a week in Paris before the Huxleys traveled south to the symposium at Le Piol in St. Paul-de-Vence. I am looking at a photograph from a few years later, a color slide taken by Osmond. Captured from Le Piol's terrace high on a hill, it shows the outdoor dining area overlooking distant countryside. I can almost smell the lavender.

What transpired here in spring of 1954 were the International Conferences of Parapsychological Studies, to use its formal title. At the first of two sessions, which began on April 19, Aldous presented a paper about the mescalin experience called "The Far Continents of the Mind." The title alluded to his metaphor of the distant antipodes, where we face unfamiliar creatures such as kangaroos, and where one oddity may be terrible and another glorious. Having set up this travelogue to the ends of the earth and noting the difficulty of getting there, he named two vehicles uniquely suited for making the journey—hypnosis and mescalin.

Writing from Canada, Humphry said he looked forward to hearing how much progress Huxley had made toward adding new names to their roster of potential initiates, and reminded Aldous that actual experience rather than just talking about it was what really counted. "My guess," he wrote, "would be that this depends upon how many of them can be lured into taking mescalin." Osmond said he hoped the philosophers at Le Piol would be "as adventurous as old Suzuki"—referring to the noted Zen scholar and Columbia Professor who lately in New York, over traditional tea at Suzuki's brownstone on West Ninety-Fourth Street, had enthusiastically volunteered to become one of the mescalinized members of Outsight. Osmond had the impression that if he'd had a supply with him, Suzuki would have undergone the initiation on the spot.[3]

Suzuki had urged moving ahead with the project and sharing results with the public as soon as possible. The first practical test of the hydrogen bomb had taken place on Bikini Atoll the month before, and Suzuki thought a life raft of testimonials about the wisdom of mescalin from the participants in Outsight could counteract North America's infatuation with weapons capable of incinerating the known world.

Osmond himself called it a matter of fighting technology with technique, of drawing attention away from the threatened "earthscape" and directing it toward "the inscape." Moreover, this would be couched in a biochemically based language science should grasp or, as Osmond explained to Aldous, "In the past there have been three ways of classifying the transcendental: the artistic, the philosophic and the religious. We now have a fourth way, the scientific."[4]

Osmond also reinforced the Zen scholar's point about the H-bomb. "The urgency of these days," he told Aldous, "makes me feel that the rather leisurely plans which we made in summer '53 should be speeded up." Their immediate goal, he said, should be to land funds for a meeting in New York. Participants would include Osmond, Huxley, Suzuki, Smythies, new recruits from among the philosophers and psychologists Aldous met that month at Le Piol, and notable names in the arts. C. D. Broad and H. H. Price were likely, too, and Einstein was, after all, nearby at Princeton.

Moreover, their case to the Foundations would be many times stronger if they had an iconic figure such as C. G. Jung on their side. The sage of Zurich had expressed his support when John Smythies met with him and had said as much again in a 1952 letter. Now a fresh opportunity might carry this further. Jung had asked Aldous to visit him and during this European trip the two were planning to meet.

The three-week stay at Le Piol had been both stimulating and restful, a chance for Aldous to air his ideas and meet remarkable people. Among those Aldous met at Eileen Garrett's symposium was someone John Smythies would want to recruit for the Outsight group: Brown University professor C. J. Ducasse. He was a member of the American Society for Psychical Research, and his campus in Rhode Island was close enough for an Outsight meeting tentatively planned for the following spring in New York.[5] They had a longer list of individuals to invite, but to launch the project it was best to start with those who could travel to New York.

After the first session at Le Piol concluded, it was followed by the "Study Group on Unorthodox Healings" between April 27 and 30, with Aldous participating but apparently not presenting. His interest in unorthodox healing dated back at least as far as his curiosity about the healer of Llano, who had been the model for the still-unpublished *Jacob's Hands.* One afternoon Aldous and Maria were invited to Eileen Garrett's private suite of rooms, where they watched a demonstration of a technique called "magnetic passes." This involved passing the hands over and around the body. Huxley was as yet unaware, but this technique would figure in the Huxleys' unfolding future.

After the back-to-back conferences ended, the couple continued on to places where postal service was so irregular that Osmond lost track of their whereabouts and became concerned. In May the Huxleys reached Egypt and spent several days in Ismailia, on the west bank of the Suez Canal, observing a doctor who often employed light hypnosis when treating patients. They visited Jerusalem, Beirut, Cyprus, and in June traveled on to Athens and Rome before returning to London via Paris.

Osmond, who had been given their itinerary and hoped his letters would catch up with his friend in London, wrote mid-June in care of Aldous's brother, Sir Julian Huxley. In this letter to Aldous he wrote, "I have had a letter from Carl Jung who seems to be keenly awaiting your arrival in Zurich to discuss the archetypal world and the need for a greater awareness of it. His letter was extremely kind and expressed 'a vivid interest' in our work, to use his own words."[6]

But the meeting with Jung was not destined to take place. Maria, feeling increasingly weak, continued to dismiss her condition as that old recurring annoyance she called by the euphemism "lumbago," but once they reached London her sister-in-law Juliette recognized Maria's deep distress and insisted that she see a physician. Confirming what Maria already suspected, that the cancer was back, the physician insisted that they return to the States without delay.

Even at this point, Aldous was unaware of the true gravity of her condition. He replied to Osmond that although Professor Jung had asked Aldous to visit, they had decided against it because Maria's doctor did not want her to stay away too long. Three weeks after Huxley's sixtieth birthday, which came at the end of July, they boarded the SS *Mauretania* for their return to New York.

CHAPTER 11

ALDOUS GOES ROGUE

T he Huxleys were still in transit when Osmond received encouraging news, except sometimes good tidings come with a catch. A few months earlier a representative from the Rockefeller Foundation had paid a visit to Weyburn Hospital, and the upshot of the visit was a grant of $115,000 on behalf of the Schizophrenia Research Group. The funding, however, was restricted to biochemical-psychological research and could not be tapped in any way for the Outsight project.[1]

"It will, however, put a seal of respectability on us," Osmond told Aldous, "and more important, help us to get more people and some essential equipment." He wrote this about a week after his thirty-seventh birthday, and added in the same letter, "I think that we may be able to use Rockefeller as a lure for the really important work very soon."[2]

If Osmond's schizophrenia group served as a lure, then of course so did Huxley's byline. *The Doors of Perception* had reached a large audience, and Osmond reported that Harvard ethnobotanist Richard Schultes had recently told him that "peyote certainly has more devotees now than ever in its history."[3]

The Outsight message needed the right kind of sustained attention if they were to land Foundation support. For his part, Aldous needed another book project. He was already working on an essay about visionary art and was scheduled to deliver a talk on this topic in Washington, DC, in September. To expand the part about visionary experience into book-length form, he needed the conceptual boost of a fresh session of mescalin. Maria, acting as liaison, passed his message to Osmond on her blue airmail stationery: Aldous had been thinking about another session, and the first little book was almost out of date. To mine and carry back material for a companion book, Aldous needed a return trip to the far continents of the mind.

Sanguine about undergoing another session, Aldous had benefitted from a positive experience with the drug a year before. Osmond was more circumspect about the risk involved every time, and knew a wrong turn when exploring the antipodes of mescalin could lead to terror rather than splendor. It had happened to the Canadian journalist Sidney Katz and to John Smythies, and these were men of disciplined mind.

Osmond had first-hand knowledge of how human chemistry could mutate into tragedy. Not long before, he had served as an expert witness in a case in which a schizoid patient in a psychotic state had killed a young girl. The memory of the gruesome act detailed in court led him to remind Huxley, "In the best of us and the worst of us, the strongest or the weakest, the winds of heaven and hell can blow with hurricane force."[4]

By now, Aldous had learned that the recent Rockefeller infusion of funding for the Schizophrenia Research Group, and a similar grant from the Commonwealth Foundation, both were restricted to biochemical-psychological research.[5] The Ford Foundation, headed by Huxley's friend Robert Hutchings, had not come up with a penny. So much for counting on connections.

Osmond, whose schizophrenia research at least enjoyed some largesse for his second-best pony, agreed that foundations maddeningly move forward inch by inch. Neither suspected that an unlikely backer of Outsight—at first glance the answer to both Aldous and Osmond's funding prayers—would soon emerge on the horizon. He was a mescalin and LSD enthusiast and a millionaire who resided on a private island in British Columbia.

By coincidence, the Pacific Northwest was in my family's sight lines that summer for a driving vacation up the coast. The purpose was to repair my parents' fraying marriage, unraveling in large part due to my father's obsession with the hand project. At some point we had a picnic lunch beside a rocky inlet and I poked around the waterline looking at shellfish while my parents caught a nap. Though I didn't know it, the view across the water was roughly toward the Dayman Island home of a man who had made a fortune in uranium and in the airline industry, and in the process cultivated friends in high places. He was Al Hubbard, the scientific director of the Uranium Corporation of British Columbia, and he liked to be called Captain. He had read *Doors of Perception*, purchased and subsequently gave away dozens of copies, and for the past year or two had been conducting his own experiments with mescalin and LSD.

At summer's end, Aldous was back on the road and presenting his series of university lectures called "Visionary Experience, Visionary Art and the Other World." Later this would be incorporated into the sequel to *Doors*, adding dimensions he had not fully explored the first time around. After Washington, DC, he lectured in North Carolina at Duke University, where the autumn weather turned unexpectedly hot, and where Huxley and Rhine agreed to drop the formalities of "doctor" and "mister" and call each other Aldous and JB.

In their conversations the main topic was ESP, or *psi*, which had been Dr. Rhine's focus for decades. Rhine had begun his research by working with trance mediums, but since then he had greatly narrowed his focus. Huxley's circle dared to entertain the idea that ESP had far greater implications, that what the mind wielded to effect ESP between rooms also came into play when stimulated by mescalin; it came into play in mysticism, and could be harnessed in healing and through hypnosis.

Rhine had looked into these sweeping areas and their possible connections in his earlier days, but now he mainly limited his work to quantifiable studies and analysis. For purposes of Aldous and JB's conversations that autumn of 1954, the topic was ESP in relation to measurable effects from psychoactive drugs, and the list of such substances was growing. Aldous learned that Rhine had been collaborating of late with a young doctor who worked with Benzedrine and barbiturates in connection with testing for ESP. Aldous shared with Rhine that his circle had been experimenting with ingesting not only mescalin but other substances, including *ololiuqui,* which was a type of morning glory vine with psychoactive seeds, used for centuries by the shamans of Mexico. A member of Huxley's Tuesday night circle, the psychologist and hypnotist Leslie LeCron, had obtained some seeds from Cuba, and he and Osmond were writing papers about ololiuqui, a potential stop on Huxley's tour of the Far Continents of the Mind.

During meetings with JB, Aldous heard about other experimental work in the field. "I learned, incidentally," he told Osmond, "that the National Institute for Mental Health is experimenting with lysergic acid—to what end I could not discover as I had no time to accept an invitation to go and see."[6]

After Aldous returned from his East Coast lecture tour, he told Osmond he wished he'd had a chance to take ololiuqui and then test the results in Rhine's ESP lab. To Osmond, pairing a simulated psychosis with an ESP experiment made sense because a schizophrenic state seemed to be associated at times with a psi state. Or as he told Huxley, "There is good evidence that early in certain schizic illnesses psi phenomena are facilitated. The highest scorer on the Zener cards I ever saw was a young man in a mild schizic episode."[7]

Huxley was keen on revisiting mescalin during Osmond's upcoming visit. As before, the psychiatrist would stay with the Huxleys; once again he would bring along mescalin, or possibly LSD; once again Osmond would administer the drug and ensure that for someone prone to health problems like Aldous this second trip would go well.

Osmond, feeling harried by a recent calendar of Weyburn obligations, asked if they might spend a few quiet days outside the city at the start of his November visit, perhaps in the desert or by the seashore, before he became a Southern California "social animal." Otherwise the latter was a certainty, because the Huxleys' friends overlapped so many spheres that if Osmond did not speak out he might be dropped straight into a maelstrom of introductions. In his reply Aldous wrote that of course he understood; he would borrow a friend's house in a place called Joshua Tree, near Palm Springs, where they could expect to enjoy the desert at its late autumn, crystalline best, and asked Osmond to please come supplied. Osmond confirmed, alluding to a 1914 act regulating opiates and hoping he would not be held up by the Harrison Act, though in 1954 mescalin and LSD were barely known investigational drugs.

Humphry's plane landed early on the morning of November 18, and he followed the same procedure as he had back in May of the previous year by taking an airport transfer to Hollywood's Roosevelt Hotel. Across the street, Marilyn Monroe's handprints, along with her high-heeled footprints, had recently been added to the forecourt of Grauman's Chinese Theater. (Two months after his previous visit, Marilyn had costarred with Jane Russell in *Gentlemen Prefer Blondes*, based on a novel by Huxley's friend Anita Loos.) From the Hollywood Roosevelt, Osmond took a cab to North Kings Road, where he and Aldous sat down to breakfast.

Mirroring Osmond's role as a tour guide to other worlds, Aldous took his houseguest on a Southern California tour juxtaposing the natural and the unnatural, high and low art. To Osmond's relief, it began with a few quiet days outside the city, where Osmond made the acquaintance of the hulking cactus called the Joshua tree. After the desert interlude with its mind-clearing vistas, Huxley added a twist: Hadn't Osmond asked for "a few days of rest"? Then Huxley would take him to the green hills of Forest Lawn, with its statuary symbolic of eternal repose.

Sightseeing included the MGM studios, before such tours were widely available commercially, MGM being the studio for which Huxley had written several screen treatments. One of Huxley's film-industry acquaintances was director George Cukor, then in the final editing of *A Star Is Born* starring Judy Garland.

During the November visit Osmond finally met Huxley's celebrated friend Gerald Heard. Osmond knew about Heard through Aldous, but also from his BBC days, his books on cultural anthropology and spirituality, and Heard's bestselling 1941 mystery novel, *A Taste for Honey*. Osmond had a consultation with Miss Hall, a fortune-teller Maria regarded as her own personal "witch." Sophia the trance medium held forth in the tiny voice that Osmond found so astonishing, purportedly issuing from a discarnate guide. He watched a demonstration of hypnosis and age regression that he would later describe to his medical colleagues back in Weyburn, making them squirm.

Shopping was again on the agenda, and while strolling down the aisles Aldous probably came up with a few choice remarks about product glut in a season when shelves brimmed with goods. Such bounty provided an opportunity for Osmond, who lived in the culturally thin hinterland, to choose holiday gifts for his wife and daughter. For Jane, he picked out an outfit he called worthy of wearing on the set of a Cukor film.

The resting and touring and shopping went well, but unfortunately the main reason for Osmond's visit, Huxley's second mescalin session, did not. This time Osmond had brought with him both mescalin and LSD. Aldous had been ill on and off since his East Coast lecture tour, which had come on top of the European trip to half a dozen countries, so his weakened condition was no surprise. But just when he was poised to renew his acquaintance with mescalin and add fresh material to his work-in-progress, Huxley's body protested with a case of shingles.

Ever since their first exchange of letters, Dr. Osmond had advised against taking mescalin if a person was prone to depression or had chronic liver problems, or if they were suffering from shingles. Maria was unwell that week, too, and Osmond called off the session. Aldous was deeply disappointed. Before Osmond said goodbye on December 3, ending any possibility of changing his mind about the mescalin, Maria drew Humphry aside. She had something to confide but he mustn't tell Aldous.[8] Yes, she was unwell, but it was not lumbago as he and her husband had been told. What she had suspected was confirmed in London that August when she had learned the cancer had returned.

Osmond left LA harboring Maria's secret and returned to Weyburn saddened and subdued. A few days later he received a letter from Maria and read it aloud to Jane, a former nurse. Though the letter seemed cheerful, he knew what Maria meant between the lines when she wrote that she and Aldous had had a wonderful time during Osmond's stay and thanked him for being so very kind.

Back home in what Jane called their "gloomy prairie," Osmond reflected on his recent trip to Southern California. "It was an extraordinary, delightful two weeks, a feast of people and places," he wrote in his thank-you note to the Huxleys, adding, "Though of course the main thing was being with you both."[9]

Once back at his desk as clinical director of the hospital, duties probably helped keep his mind off the truth Maria had entrusted to him, though with this news had come a foreboding of the profound grief that lay ahead for his friend. Still, it was the season of festivities and relentless well-wishing. Game to provoke colleagues and play the gadfly, Osmond found it amusing at holiday gatherings to recount how, while a guest of Aldous Huxley, he had watched a demonstration of regression to previous lives. The majority of Osmond's colleagues were disturbed by the very idea of such reincarnation nonsense, much less hearing about a demonstration hosted by someone much admired like Huxley—a demonstration condoned by their science-grounded colleague Dr. Osmond. The critics argued that even if a few facts conjured up through so-called age-regression hypnosis could be verified through records, this remained purely anecdotal.

Afterward, Osmond reported to Aldous, "One of my doctors, an intelligent man, even insists that science only concern itself with what can be easily repeated. It is hard to see how astronomy, archaeology, sociology, geology would be judged if this were the criterion of science. There is a tendency to equate the controlled experiment with science although this is only one of many methods which are available."[10]

In snowy Saskatchewan that Christmas Eve of 1954, Osmond sat down to write a note to Aldous and wryly observed, "Weyburn has been full of various sorts of rejoicing, including the very busy liquor store." Gifts brought back from California (Jane's new attire, little Tig's cowgirl outfit, plus a doll from Maria's sister Rose) were arrayed under a bedecked tree in the Osmond's home.

In our cottage in Southern California, where a row of holiday cards sat on the mantelpiece, the one on creamy thick paper with an unusual black-and-white woodcut design stood out from the rest. Inside, the note in Maria's fantastical script read, "Wishing all the best to the three of you, from your friends the Huxleys."

I wonder what would have happened if Huxley had not suffered the attack of shingles in November of 1954 so he and Osmond could have gone ahead with their plans. The Canadian psychiatrist would have supervised and recorded impressions, while Aldous underwent his mescalin—or perhaps LSD—trip-to-the antipodes for the second time. By January, Aldous would have been hard at work on his new book, and he might have said no to a risky invitation when Captain Al Hubbard showed up with a considerable cache of drugs.

More importantly, I think, Maria might have been spared twenty-four-hours' worth of worry in the last few weeks of her fragile life. I mentioned that Al Hubbard liked to be called the Captain, and that gives you some idea of what Maria would soon perceive about this man—that he intended to step in as a new boss in the absence of Dr. Osmond.

John Smythies, now doing research in Vancouver, had set events

in motion the previous fall when he had met the millionaire, one of
wh[...] homes was an estate on a private island in British Columbia.
Sc[...] . Similar to the way
G[...] referred people of
"si[...] the same. Moreover,
Sr[...] ke a way to obtain
fu[...] c project based on
in[...] l conveying how this
tr[...] nankind.
[...] d—a promising and
in[...] d, moreover, was so
en[...] opies so far. Osmond
sa[...] efeller as among his
in[...] he would be glad to
go[...] ange funding for the
O[...] hought could reason-
ab[...] ary.
M[...] ar now added Gardner
[...] nerican Psychological
As[...] iversity of Chicago, to
name a few. From the initial [...] pected more invitations
would spread through the network to contacts of their contacts.

"It will be one of the most exciting pieces of work ever done,"
Osmond told Aldous. "Do you know of any occasion reported in
which many really able people have been able to meet and discuss
transcendental experience from their own experience? I am very
hopeful that if we pick the right people we can change the intellectual and spiritual climate very quickly."[11]

On January 11, 1955, Aldous wrote, "Dear John, Your friend
has turned up here," meaning the Captain, who had arrived in Los
Angeles at what seemed an opportune time. It had been almost a
year since *The Doors of Perception* hit the bookstores. *Life* and *Esquire*

were courting Aldous with attractive fees to write essays, and a flow of articles needed a flow of ideas.

Aldous invited the Captain to North Kings Road for a luncheon also attended by Gerald Heard. Maria had reported on January 4 that the Captain was a success and Aldous found him very clever. Apparently so did Gerald, for that same week the Captain went back and forth between his hotel, Gerald's place in Santa Monica, and Huxley's home. At some point, the Captain and his wife were invited to a séance with Sophia, except that particular evening Sophia was ill and did not show.

Maria had noticed something about the Captain during the luncheon earlier that week, having to do with a configuration she saw in his hands. Her own hand studies, and the prints she had collected for years, had been based on theories of Charlotte Wolff, with whom she and Aldous had studied for a while in Sanary during the early 1930s. In Hubbard's hand she saw the signs of a person who could be swept away by enthusiasm.

Hubbard would later be credited with introducing new techniques in LSD-therapy, such as eliciting buried memories and experimenting with what came to be called "set and setting," whereby he modulated the room for a session through use of lighting and music. But as Erika Dyck notes, Hubbard would pursue his LSD network so aggressively that, lacking medical or scientific credentials, he eventually had to source his LSD from questionable suppliers.[12]

But that was still in the future on this cusp of the new year of 1955. The relationship with the Captain seemed promising, and, by the end of the year, Osmond, Smythies, Huxley, and others would have lent their names to a new network under Hubbard, called the Commission for the Study of Creative Imagination.

The day after Sophia's séance was cancelled, Gerald telephoned Aldous with a tempting offer. The Captain was presently at Gerald's in Santa Monica, and the two men were planning to take mescalin. It would be without the presence of a physician. This was Hubbard's way.

Aldous, who by now had recovered from his earlier bout with shingles, was aware of what could go wrong, and he knew Gerald suffered from recurrent depression, which was one of the conditions Osmond advised against mixing with these substances. Still, the Captain had the goods. Would Aldous like to come? Of course he would.

As I understand it from her January 9 letter to Osmond, Maria did not want Aldous to participate without a doctor who understood the risks of the drug and how to administer an antidote. If the men insisted on taking it unsupervised, then she wanted them to do so at the Huxleys' house, where Aldous (who sometimes used his fingertips to skim the walls when he navigated between rooms) would have Maria nearby if he needed her. But it was nothing doing, as Maria later told Osmond. It would be at Gerald's house with or without Aldous, who decided it would be with.

Inside Gerald's house in Santa Monica, refreshed by breezes from the nearby sea, four men gathered for the experiment. One was a young philosopher about whom I have no details. The others were the Captain, Heard, and Huxley. I picture a tall, thin Aldous clad in pleated tan trousers and a white tailored shirt, with no tie but perhaps a vest and tweed jacket because he was thin and January days in coastal California turn cool. The thick thatch of dark hair on Huxley's leonine head was, by now at age sixty-one, mixed with strands of grey.

I see Gerald Heard, shorter than Aldous and slight of frame, with his brown-grey beard in a modified vandyke. He is also wearing a long-sleeved shirt, a jacket, and no tie. Unlike Gerald and Aldous, the Captain is stocky, and I suspect he doffs his jacket indoors because this climate is milder than what he is accustomed to in British Columbia. The four settle in around a coffee table, where the Captain doles out the dosage for all and they wait for it to take effect.

The Captain explains his plan: to see what happens in group mescalinization, an experiment in social interaction rather than the usually solitary experience. For his part, Gerald frames it as changing the focal length of awareness to a wider angle. Aldous and Gerald are probably wondering if the luminous but lonely experience extends to shared insights. The Captain reveals his underlying plan—he sees this session as quasi-confessional, a device for raising buried guilts and reliving traumas. In the end, Aldous later told Osmond, the group mescalinization experiment worked better for Gerald and Aldous than it did for the Captain. Afterward, Aldous spent the night at Gerald's. It seems the dosage was so potent that none of them were capable of driving him home.

The session did prove fruitful, as Aldous later acknowledged to Osmond: "Gerald and I . . . went somewhere else—but not to remote Other Worlds of the previous experiments . . . it was a transcendental experience within this world and with human references."[13] The new material would be incorporated into the sequel. Huxley was almost ready to pull together his next book, one about the visionary experience.

Even if Maria had misgivings about the Captain, Aldous and Osmond saw him as a benefactor. "I am hopeful," Aldous told Osmond in his January 12 letter, "that the good Captain, whose connections with Uranium seem to serve as a passport into the most exalted spheres of government, business and ecclesiastical polity, is about to take off for New York, where I hope he will storm the United Nations, take Nelson Rockefeller for a ride to Heaven and return with millions of dollars."

Though she had objected to Aldous attending what she considered a risky mescalin session, Maria was perhaps consoled to learn afterward that medical advice was available, had it been needed. At some point while at Gerald's, Aldous did, in fact, place a long-distance telephone call and speak to Dr. Osmond in Canada.

CHAPTER 12

MARIA

The previous summer, after their stay at Eileen Garrett's Provençal estate for the second Le Piol session, called "Unorthodox Healings," the Huxleys had traveled to Cairo where they observed a medical treatment in which patients were left in a hypnotic sleep state for a period often spanning many days.

Hypnosis (a word derived from the Greek word for sleep) was a frequent topic for Huxley's Tuesday circle, where experiments were often presided over by psychologist and hypnotist Leslie LeCron. In Huxley's yet-to-be-written utopian novel, hypnosis would be central in medicine; in an earlier work, *The Devils of Loudun,* Aldous proposed group hypnotic suggestion as a way to account for the so-called diabolically possessed Ursuline nuns.

The National Guild of Hypnotists was founded in the United States in 1951. In the early 50s it was common for hypnotists to practice age-regression hypnosis, probing in a way Freud might have approved of to uncover early life traumas below the threshold of conscious memory. Hypnosis made headlines in 1952 when an amateur hypnotist named Morey Bernstein was doing a standard regression on a Colorado housewife named Virginia Tighe and asking questions about her early childhood—except when Bernstein prodded Virginia she kept going back further, until her voice changed into an Irish brogue and she said her name was Bridey Murphy. When asked for details about her life, which were tape recorded by Bernstein, many details seemed to align with records from a century ago.

The improbable idea of age-regression to a time before birth requires an extraordinary explanation. One possibility is that an individual alive today is the reincarnation of someone who has lived before, and hypnosis uncovers a deeply buried memory of a former life. In a variant, an individual taps into the disembodied memory or imprint of someone no longer walking this earth, which might also account for a change in a trance medium's voice.

The story of the Colorado housewife simmered on in the newspapers until Bernstein's book, *The Search for Bridey Murphy*, was published in 1956, when it became a bestseller. Once the full details of the story, or perhaps elaborations, were fixed in type, journalists and debunkers began scrutinizing the particulars. It turned out that although some details recounted under hypnosis could not be explained away, others could—they might have been stories Virginia overheard during childhood from her Irish immigrant neighbors. Plausible explanations weakened Bernstein's claim, and the Bridey fervor died down because doubts were cast or because fads just fade away.

Not long after hypnotist Leslie LeCron and Humphry Osmond met for the first time at Huxley's in May of 1953, the two men developed a mutual interest in the anthropology of hallucinogens and their use in sacred rites in Mexico and South America. The cactus peyote and similar substances had been used for centuries to access or interact with spirits believed to inhabit the Other World. The boundary between the mundane world and this Other World becomes permeable on a shamanic journey undertaken by ingesting sacred substances in a specified ritual.

For LeCron and the Tuesday circle, experiments with hypnosis raised questions about other dimensions and the possibility of personal survival after death. Maria had once pointed out that in the state induced by taking mescalin, unlike the state experienced by a schizophrenic patient, one can take comfort in knowing one's mind will return to normal, but one does not return to normal after the final journey across the causeway between worlds.

In early January, just after the session with Captain Al Hubbard, Aldous was still buying into Maria's excuses. In his January 16 letter to Osmond, Aldous reported that Maria still had lumbago, which meant in theory that the Outsight meeting in New York might have gone forward that spring. Captain Al seemed confident.

Had it happened, Osmond and Smythies and Hoffer and Huxley would be there, and at this point Captain Al, too, and probably Gardner Murphy, former president of the American Psychological Association, and psychologist Heinrich Kluver of the University of Chicago, who had done early work with mescalin. Certainly professor Deisetz Teitaro Suzuki, the leading interpreter of Zen Buddhism in the West, and perhaps novelist Graham Greene, as well as philosophers John Curt Ducasse, Gilbert Ryle, H. H. Price, Charlie Broad, and logician Alfred Jules Ayer. Professor Carl Jung would be invited, as would Albert Einstein, who was at the Institute for Advanced Study in Princeton.

Of course this scenario didn't happen, but, if it had, here is a possible alternate history: A high-profile Outsight gathering in 1955 might have ushered in well-funded studies, and that might have changed the plans five years later of a young professor who came to Harvard and sent psychedelic research off the rails.

The main reason such a meeting did not take place was that, in late January, Aldous learned that his wife's cancer was terminal.

My father was shocked when he learned that Maria had at most a few weeks to live. How could he not have noticed she was that ill? Are some women so skilled, he wondered, at making themselves look healthy when they are not? Of course some men are more oblivious than others, and especially those who, like Howard and Aldous, live mainly inside their own heads.

Dad knew my mother would be upset because she was fond of Maria. He cared for Maria, too, but his reaction when he realized he would not see her again was a rush of guilt. When Maria had first seen his hand photographs with their remarkable detail she had asked his permission to have someone copy the custom camera and build one for her.

Not being proprietary about his design, he had gladly agreed. "The next time I went to the Huxleys," Dad said, "Maria was almost in tears. She said she couldn't find anyone to build it because they said in order to understand how it worked they would have to take my original apart." His camera was often in use, and at the very least a dismantling would delay his scheduled appointments, including meetings with a UCLA psychology professor who said he was interested in replicating the hand study.

Maria asked my father if he would be willing to construct a copy of the camera and said that she was willing to pay whatever he thought fair.

"I said I was very sorry but I just didn't have the time," Dad said. "I've always regretted it. I had no idea she was dying."

The Tuesday night circle would quietly disband; it would no longer meet for dinner at the World's Biggest Drugstore then regroup in the music room of the house on North Kings Road.

Maria was hospitalized on January 22, and, though technicians administered radiation in a futile attempt to reduce the pain in her spine, nothing more could be done. She was brought home and all engagements were cancelled.

Aldous wrote a note to Osmond, perhaps as correspondence therapy. Maria had lost consciousness. Though Aldous spoke softly of the visionary world and the Light beyond it, "it is hard to tell if she hears me, but I hope and think that something goes through the intervening barrier of disintegration and mental confusion. The

Bardo Thodol," the *Tibetan Book of the Dead*, "maintains that something penetrates even after death."[1]

Weyburn was wrapped in a cold spell toggling between twenty and thirty degrees below freezing and a minor blizzard was underway around the time a large manila envelope arrived from Los Angeles. Inside, Humphry found a six-page typescript, a singular installment in the decade of letters and documents exchanged between the two men. This was Huxley's account of Maria's death, and the accompanying letter was mistakenly dated February 22, 1954, as if Aldous had willed one more year.

Osmond felt honored to be included. "I am very glad and grateful that you told me about the last days of someone I deeply admired and loved," he told Aldous. "How little attention doctors pay to death. . . . It is just a way across the border, an easing of the strands that bind us to this world and which separates the two."[2]

Matthew came from New York to be with his father, and, during the following weeks, Aldous maintained a shadowy semblance of daily life with help from a housekeeper and his loyal cook. He wrote Osmond that though his health was fair, "the house is full of the presence of an absence."[3]

After a respectful period of mourning, friends tried to interest him in activities. By mid-March, with Aldous still struggling to adjust, some suggested the tonic of resuming a public schedule, and accordingly he decided to keep a previous commitment to give a talk in Princeton that coming May.

After discussion, a plan took shape. In April he would travel across the country with his sister-in-law Rose behind the wheel. It would be a twelve-day, four-thousand-mile journey, swinging southwest so Aldous could visit Frieda Lawrence, D. H. Lawrence's widow. Afterward, they would continue to New York, where Rose would visit friends, and for

Huxley it would be on to Princeton, then an extended stay with his son's family at their Connecticut country home.

Some thought the two-week excursion was a compatibility test for Aldous and Rose. At least the journey would provide diversion for Aldous, who had enjoyed road trips ever since he and Maria navigated European roads in their scarlet Bugatti, anticipating curves and relishing the rush of speed. By June, the road trip and visit to Frieda and the conference was over. Aldous was resting and writing under the trees at the home of his son, Matthew, daughter-in-law, Ellen, and grandchildren, Tessa and Mark. During those summer days, he finished the draft of *Heaven and Hell*. Friends who fancied themselves matchmakers, who maybe were betting on the road trip with Rose to launch a deeper relationship, would be proven wrong. Someone else was waiting offstage.

CHAPTER 13

CONSTERNATION AT MENNINGER AND MIRAMAR

After Maria's death the lives of my father and Humphry Osmond took oddly similar turns. Chances are they only met once face-to-face, and that was at the Huxleys' home in May of 1953, but they had in common a Huxley-related, innovative, and risk-taking bent. This led to both men presenting their findings before an audience with the power to make or break them. For Osmond, this occurred in 1955.

Osmond attended a conference on the East Coast (not the same one Aldous would be attending that spring), and before returning to Weyburn he added two stops. In New York he met with Bill Wilson, the cofounder of Alcoholics Anonymous, where during their meeting he told Wilson about his and Hoffer's recent experiments administering mescalin or LSD to a small sample of alcoholic patients. Afterward he described this meeting to Aldous, saying he found Wilson "very lively and keen on our new idea for using mescalin etc. as a transcendent experience for alcoholics."[1] Wilson would first try LSD a year or so later.

What led to Osmond and Hoffer's eventual LSD treatment for chronic alcoholics could be called off-label. As intended by its manufacturer Sandoz when released as Delysid, LSD produced a model psychosis and also had potential as an aid in psychotherapy. In its role as simulating psychosis as an experiment, some nursing staff at the hospital in Weyburn eventually underwent mescalin or LSD sessions to better understand their schizophrenic patients.

In the early 1950s, Osmond and Abe Hoffer began exploring the therapeutic potential of both drugs, eventually narrowing it to LSD. After exploring hallucinogens for schizophrenia, one of the next areas of experimentation was alcoholism. Because a hallucinogen such as mescalin or LSD simulates psychosis, Osmond and Hoffer wondered if this drug might simulate other medical conditions that, under study, might lead to more effective treatments. One condition the two doctors considered was *delirium tremens*, or the so-called DTs, triggered by severe alcohol withdrawal, which resembles a transiting psychosis often accompanied by hallucinations (giving rise to the cliché about seeing pink elephants).

But the idea of simulating *delirium tremens* was shelved in favor of another LSD research direction: Exploring why some patients claim to have entered an ecstatic state during alcohol intoxication. Huxley suggested to Osmond that he investigate this phenomena. He wrote to him shortly after *The Doors of Perception* was published in February of 1954, when letters from readers all around the nation started arriving at Huxley's home (presumably through his publisher) shortly thereafter.

In a letter to Osmond dated March 2, 1954, Huxley wrote about one letter he had received: "Another stranger writes from Los Angeles. He is an ex-alcoholic who had ecstatic experiences in his early days of alcoholism and insists, in spite of what the Freudians say, that the longing for ecstasy is a very strong motive in many alcoholics. He is also a friend of Indians, knows some who have taken peyote. . . ." A few lines later, Huxley added, "He suggests that it might be very interesting to try the effects of mescalin on alcoholics past and present. And I think that, if your research project gets started (or even if it doesn't), this might be a fruitful thing to do."

Osmond and Hoffer did find this idea fruitful, and they developed a technique for inducing a visionary experience, an epiphany of sorts, in alcohol-addicted patients. The goal was for a patient, who was given LSD in a medically controlled setting, to encounter an ecstatic or extraordinary state of mind unrelated to alcohol. This model of alcohol therapy would involve a single large dose of LSD to induce a psychospiritual or transcendent breakthrough. Osmond and Hoffer believed this experience provided the best chance of breaking the chain of alcohol dependency, and because it involved a one-time dose of LSD, a substance not shown to be addictive, it was not likely to lead to further addiction.

After the meeting in New York with Bill Wilson, Osmond's next stop meant another chance to land the kind of endorsement and support for Outsight he had been hoping for. This meeting took place at the Menninger Foundation in Topeka, Kansas, with Karl Menninger, who along with his father, Charles, had founded the Menninger Clinic in 1925, and established the Menninger Foundation in 1941. Osmond hoped to propose three projects to Dr. Menninger. One was the use of LSD to help cure intractable alcoholics, an approach that later came to be known as the psychedelic treatment model. At this point in time Osmond did not realize that his work with alcoholics would turn out to be his proudest legacy.

The second project Osmond hoped to propose to Menninger was Weyburn's "schizophrenia-mescalin etc." study, which, though Rockefeller and Commonwealth foundations had finally provided initial support,[2] still needed funding if the schizophrenia group in Saskatchewan was ever to find the elusive biochemical link accounting for the resemblance between mescalin and adrenaline.

Like a pyramid of projects, the pinnacle was their prospectus describing Outsight, or what Osmond called his "best pony," but I think their highest cause was how Aldous saw it. Outsight had the potential to improve countless lives, and perhaps through diffusion of transcendent awareness even slow down what seemed to be acceleration toward nuclear destruction. But this was the most difficult of the three proposals to describe. Outsight's pitch—when he had a chance to make it a few months later—would start with the celebrity name of Aldous Huxley, who would participate in a steering committee, which in turn would invite notable individuals to participate, after which Doctors Osmond and/or Smythies would administer mescalin or LSD, record subjects' impressions, glean insights, and propose a program for wider application. A logical order.

They had it all worked out in eight points on paper.

I imagine Osmond was thinking how it would be so much easier if Karl Menninger would just take the stuff and see for himself.

The first brief meeting with Menninger went well, and although Osmond had a head cold on the way to Topeka, it barely dampened his enthusiasm. As he wrote to Aldous in mid-April of 1955, "This is one of the greatest psychiatric centers in the U.S. and to capture it for the new approach"—that is, the biochemical approach to schizophrenia—"or at least to make them reasonably interested would be a triumph, but of course exactly the opposite might happen."

Afterward he reported that Menninger was not only fascinated by the work he was doing in Canada but "even more so by our work together," meaning Osmond and Aldous jointly, and that fascination seemed to bode well for the top-tier project he hoped to propose.

After their first visit, Menninger sent Osmond the latest edition of his book *The Human Mind*, personalizing it with the date April 1955, "For Dr. Humphry Osmond, who excited us all to new thoughts and visions!"

Menninger invited Osmond to return to the clinic, and to Humphry's amazement he was asked if Osmond might consider joining Menninger's staff. "He indicated that I had only to name my price," Osmond told Huxley, asking for advice: Should he ask for double the salary he currently received in Weyburn or should he ask for more?[3] The prospect of a career move to the States had a definite appeal, as it meant a chance to relocate his family away from the Saskatchewan hinterlands his wife so greatly disliked and that they called "the dreary prairie."

He was buoyed by the open-ended job offer and still basking in reflected fame since publication of *Doors of Perception*. On this second visit, Osmond wrote Aldous that he would be "seeing whether LSD will help a few very intractable alcoholics. I think it may and so does Bill Wilson"[4] Osmond scheduled the return trip for the end of May.

It may be revealing of Osmond's state of mind, in a foreboding kind of way, to know that in the days before returning to Topeka his reading choice, as he told Aldous, was Evelyn Underhill's *Mysticism*. This hints that he had set his sights on the Very Big Picture, for once you've glimpsed the Ground of All Being it is hard to dial back your expectations.

Now it was May, and, as I reconstruct this incident from his letter to Aldous, I picture Osmond arriving at the Menninger Clinic, where it has been agreed in advance that he will work with one or more of Menninger's intractable alcoholic patients.

Osmond worked all day Thursday with a patient who was not only an alcoholic but also "a very fascinating multiple addict," he later told Aldous. It happened that Huxley and Osmond shared an interest in poetry. This addict had translated the work of nineteenth-century Symbolist poet Charles Baudelaire, who experimented with hashish and was best known for his collection *Les Fleurs du Mal* (Flowers of Evil).

Before beginning that day's work, Humphry skipped breakfast. Then rather than interrupt the flow of his ministrations to the fascinating multiple-addict, he also skipped lunch. "We were to go out to an early supper," Osmond told Aldous—but first came an unscheduled cocktail hour involving "a couple of long whiskies."

Quite famished after almost twenty-four hours with no food, and trying to manage the effects of two drinks on an empty stomach, Osmond couldn't quite figure out when dinner would be served, though, with a rising level of alarm, he hoped it would be soon. Finally, Menninger's party set out for the country club.

At last they were seated. All along, the impromptu schedule had suggested a social evening in the presence of other colleagues. Therefore it came as a shock when all of a sudden Menninger blind-sided Osmond by getting down to business. He directed a critical question to Osmond: "What would you do if you had a free hand?"

Now this might refer to working with intractable alcoholics. It might refer to working with schizophrenic patients. It might mean both. But by this point in the evening Osmond took "free hand" as a signal to trot out his best pony. "I told him. It shook him," Osmond told Aldous.

It seemed that without inhibition, thanks to hunger and whisky, and much as if speaking with Aldous in one of their free-wheeling moments, "I told them that we were faced with devising a new society. My friend Derrik Miller, a very diplomatic young Englishman was horrified."[5]

He went on to laud the eight-point Outsight program for the benefit of those assembled, and boldly pegged its cost as "work I might possibly do to the tune of, I suppose, at least $1 million," a serious sum in 1955.

Possibly it was all a misunderstanding. Or maybe Menninger wanted to challenge him with "if you had a free hand," but either way Osmond misjudged the moment. As he later told Aldous, "I was foolish enough to accept his request. It was horrid of me and in a way I went berserk. No jumping on the table . . . but the consternation was great."

The dinner performance apparently led Menninger to quietly drop the offer of a position in Topeka, and funding dried up before it could begin to flow. Yet Humphry seemed to take this all in stride. "Dr. Karl is gifted and very able, but not I fear very far sighted," he told Aldous. "He doesn't read enough science fiction."

Huxley summed up the situation this way: "I suppose it was more than could be hoped—that Menninger should be simultaneously the fountain-head of American psychiatry *and* completely open to new, revolutionary ideas."[6]

Since the publication of *Doors*, Osmond had found himself riding along on something of a celebrity carpet. To some extent my father did, too. After Dad had been attending Huxley's Tuesday night salons for a year or so, his hand project came to the attention of UCLA Psychology Department Chairman Dr. Joseph Gengerelli, called Ginger by friends and close colleagues.

The professor designed his own version of the study to see if he could replicate Dad's results, and it turned out that Ginger's results matched.

The Gengerelli-Thrasher study took place over the course of several months. My father had collected hand photographs of two groups of patients diagnosed with schizophrenia, one group at Metropolitan Hospital in Norwalk and the other at Pacific Colony in Pomona. His so-called normals had come from groups ranging from attendees at the Long Beach Hobby Show to musicians and insurance salesmen.

In Gengerelli's replication study, the schizophrenic group consisted of one hundred patients from the VA Brentwood Hospital, Veterans Medical Center–Sawtelle facility, near UCLA. The control group included the Westwood Kiwanis.[7]

One day when we were sitting in the patio, I asked my dad how Gengerelli went about the process of replication. "Ginger took photos and used my equipment," Dad said, "but UCLA wouldn't finance the cost of film so I gave him the money to buy it. I didn't actually buy the film myself because it was important that I not touch it. I told them what to get, how to process the film. I wrote it all down."

My father recalls that the control group came short of the requisite number, so Ginger threw in a few students to bring the total up to one hundred subjects. "The other hundred, those in the VA

Hospital, were quite disturbed," my father said. "They would rub food in their faces and we really had a time with those, but we wound up getting one hundred, and his curve fell right on mine."

I know that over the years a number of breakthroughs have come from amateurs, but this was a different story. I asked Dad why an academic would become interested in an amateur's work in the first place. "Because he looked at what I had already done—the photographs, the data, the instruments—and he could see it," Dad said. "He saw in the photographs and in the data that these hands were much different than the others. I could draw a curve as to what they were."

Dad said he thought he was revealing something important, something professionals could build on. "If I hadn't had the backup of Norwalk and Pomona and hundreds of hands, then who would have started from scratch? I had a solid base and good names in back of me." Those names, of course, included Aldous Huxley.

The Gengerelli-Thrasher hand study of the grain pattern—with its key feature a possible physical marker for schizophrenia—would make its professional debut in December 1956 at Santa Monica's Miramar Hotel, site of the fifty-fifth annual meeting of the American Anthropological Association.

I found a copy of the 1956 brochure in one of the boxes in the garage. That same year, Sigmund Freud's portrait graced the cover of *Time* magazine.

I am looking at the conference brochure from that December meeting. I open the program and try to picture what it must have been like on that day. For my father, hopes were high. This was a culmination of many years' work.

To reconstruct the scene from the program and my father's recollection, it was about 1:30 p.m., and a paper was about to be copresented in the Garden Room by Dr. J. A. Gengerelli of the University of California, Los Angeles, and Mr. Howard Thrasher of

North American Aircraft. The talk was listed as "Hands and Psyche: A Tentative Empirical Inquiry." Dad and Ginger had agreed in advance that the psychology professor would do all the talking.

"When Ginger threw his picture of a hand on the screen many people left," Dad tells me, "but there was another lecture going on in the building, and when it ended people came in from that lecture and saw something interesting was happening here. They filled the seats and stood in the aisles."

I ask what happened next.

"Well, someone got up and said something like, 'What do you think you're doing? We have proven beyond doubt, and the whole of the psychology community has agreed, man is the result of his environment. You're bringing it back to heredity again? We've been there. Shut this thing down.'"

"What! You mean someone actually wanted to *shut it down*, right in the middle of the presentation?"

"I'm only remembering the gist of what was said but, yes, someone used that phrase," he says. I am surprised but ask him to go on.

"They chewed on Gengerelli, but he was one tough man and he stood his ground. He chewed back. He was mister statistics, and he said to the hecklers, 'I'm staking my reputation on this. You don't know what you're talking about.'"

Apparently the objections didn't stop with the Freudian-Behaviorist offensive. Some people flat out ridiculed the resemblance to palmistry, even though Gengerelli was not talking about lines or fingers or palms but the configuration of ridges, or what is often called dermatoglyphics and what my father simply called "the grain."

"Gengerelli's only interest was the grain pattern. It isolated that one group of people." Which is to say those with schizophrenia, the pattern apparently found in the hands of patients from the VA Hospital. "Only one in the one hundred control sample had it," Dad says. "That by itself made people at the scientific meeting so damned mad, and that was because Ginger had some proof it wasn't only about environment. There's something else going on."

After the session, Gengerelli announced that he would be in a side room and available for anyone who wished to speak with him. Dad recalls one retired psychologist who came in, shook Ginger's hand and said something like, "It takes a strong man to do what you just did." Ginger's resolute stand at the conference led my father to believe that statistics would triumph, that they would fight the good fight, but afterward when the two of them were walking back to the parking lot he noticed that Ginger looked downcast. "The reaction was more vitriolic than he'd anticipated," my father recalls.

Ginger told him that no matter how carefully the data had been collected and analyzed, the kind of ridicule they had just witnessed would overwhelm their findings. *Maybe the study comes too early. Maybe someday things will change,* Ginger said, or something to that effect, but with his career at risk he said he could not stay involved. He was pulling out, and advised my father to do the same.

First stunned, then angry, my father reached his Studebaker in the parking lot, opened the door, and shoved a dozen of his best photographs under the seat. He says he doesn't remember driving home from Santa Monica to Long Beach, though in closing the distance on those thirty-two miles he had time for bitter reflection. Over the past seven years he had photographed and measured one thousand hands, including the hands of engineers and stutterers, insurance salesmen and musicians, trance mediums and twins and patients suffering from schizophrenia, and many of the members of his own family, members of Huxley's family, and the hands of Huxley's experimental circle of friends. And now, in one afternoon, it was over.

The day after the conference he sealed up, mummy-like, in boxes swaddled in silver tape, all materials related to the hand project, anything related to Huxley's Tuesday night circle disbanded two years ago, prior to Maria's death, and everything related to St. Francis-by-the-Sea and his previous work with healing. He wedged the boxes into a gloomy back corner of his father-in-law's garage. He was done with it, done with all of it, or so he thought.

Photographs from the presentation at the Miramar Hotel had a different fate. When he sold the Studebaker months later, the prints drove away with the car, forgotten beneath the front seat, but the original slides remained, and still remain, intact.

CHAPTER 14

SUCCESSOR WIFE

O nly half in jest, Aldous had called Karl Menninger someone "not open to new, revolutionary ideas" after Osmond's disappointing Topeka incident. That was when Aldous was still at loose ends after the death of Maria. He had been trying to finish a book, not ready to move forward, but four or five months later he had begun to entertain a few new, revolutionary ideas of his own.

For more than three decades, Maria had been the partner who managed their private and public life, not to mention reading aloud and processing manuscripts and coordinating his professional calendar and minding his health. Now Huxley would court a different type of woman, who orbited outside the circle of his family and friends.

Laura Archera, born in Italy and eighteen years younger than Aldous, originally came to the United States as a violin prodigy. Prevented by World War II from returning home, she joined the Los Angeles Philharmonic Orchestra, where performances took place in the downtown Philharmonic Auditorium near the grand Biltmore Hotel. Once celebrated as the largest concrete building west of Chicago, the imposing auditorium had been updated with a Deco Moderne facade and a stage with nested arcs resembling the Hollywood Bowl a few years before Laura took her chair in the orchestra.

Laura's hand just missed being included in my father's collection. By the time Dad captured photographs of most of the members of the Los Angeles Philharmonic, which must have been around

1950, she had already moved on and was working as a film editor, before making a later career change to become a psychotherapist. Aldous met Laura in the late 1940s, when she drove to the house in Wrightwood seeking his advice about a film project she and a friend were developing. That project concerned a famous Italian horse race, and her friend was Virginia Pfeiffer, a former sister-in-law of Ernest Hemingway.[1]

Laura did not figure as much in the Huxleys' world as you might say she passed through it. Laura and the Huxleys saw each other socially in Los Angeles in 1952 and in Italy during the summer of 1954, and Maria is said to have considered Laura as "a friend with similar interests."[2] And that is the point where things get interesting, because reportedly when Maria understood the gravity of her illness she singled out Laura and bestowed an implied blessing. Huxley's biographer Sybille Bedford describes a certain evening when Maria took Laura aside for a talk, but it is doubtful if Aldous was aware of any such intent because he was still unaware of Maria's condition.[3]

We do know from Bedford that Laura's name was listed in Aldous's address book in April of 1955, at the time of the cross-country road trip accompanied by Maria's sister Rose.[4] We do know, after he returned from his son's home in Connecticut, he and Laura quietly began seeing each other, occasionally sharing dinners at North Kings Road and spending more time together. Apparently no one noticed, except maybe the cook, who left prepared dishes for him and thought that Aldous's appetite certainly had improved.

Huxley's courtship of Laura Archera was so private and almost clandestine that Aldous seemed to have forgotten that he was a public figure. He assumed the ceremony would be a quiet affair when he and Laura were married on March 19, 1956, at a quickie wedding chapel in Yuma, Arizona—except reporters keep watch for such sly sacraments and the Huxley nuptials made international news.[5] Headlines reached Connecticut before Aldous could even tell his son about the marriage, which took place one year and one month after Maria's death.

Second wives seem to prompt comparisons with their predecessors, and these two women had a number of qualities in common. Both had a stylish sensibility, a European flair, and were multilingual, and both retained charming traces of their respective Belgian and Italian accents. On the other side of the list, apart from Laura being a dozen years younger than her predecessor, Maria had been constantly attentive to Aldous. Laura, as a previously unmarried career woman, was not so inclined, and in coming years she would follow her own pursuits, for though the two would live together and support each other in many ways, they often kept separate work and travel schedules. Apparently Aldous was willing, maybe somewhat eager, to take on more independence.

With its thick walls and shaded interior, the Spanish-style home on North Kings Road had served as Aldous and Maria's address for the previous half dozen years. Aldous had told Osmond a year earlier that he was aware of "the presence of an absence"—not in the sense of a haunted house as much as a house haunted by memories.

The new couple started their marriage at the house on North Kings Road, but these walls had already seen too much history. In July 1956 they moved to a modern home on Deronda Drive at the top of Beachwood Canyon in the Hollywood Hills, not coincidentally near the home of Laura's longtime friend Virginia Pfeiffer. I think it is worth noting that just as Aldous and Maria had based their relationship on an open marriage, one in which both Aldous and Maria appreciated women, so did Aldous and Laura, and her close relationship with Jinny Pfeiffer would endure until Jinny's death.

The Deronda Drive home, only six minutes from the heart of Hollywood, overlooked a canyon dense with the foliage that appears so delightfully rustic but periodically incinerates in California's fast-moving brush fires. Standing on a bluff, the modern house naturally caught sunlight, an effect amplified when Laura installed white carpets for an effect almost as bright as if Aldous was back in the desert light of Llano. This was an improvement over the dark house on North Kings, for natural desert light had been beneficial for his vision and productive for his writing.

Now that he was settling into their new home, conferences and speaking engagements loomed. The "mescalin etc." problem remained. Osmond and Huxley scratched their heads over what to do about it.

CHAPTER 15

TAKE A PINCH OF PSYCHEDELIC

*H*eaven *and Hell* arrived on bookstore shelves in February 1956. By the time of his marriage the following month, Huxley's calendar was already speckled with speaking engagements. This reminded him that he and Osmond had some unfinished business to take care of resolving the "mescalin etc." problem.

The two had often referred to this pesky topic. For three years now they'd had to resort to the catch-all phrase "mescalin etc." they coined in 1953, but it had outgrown its usefulness. The makeshift term now stood in for a lengthening list of psychoactive substances both natural and synthesized. These ranged from the peyote-derivative mescalin to LSD-25 to the mushroom-derivative psilocybin, as well as the seeds of the flowering morning glory called *ololiuqui* and later the Amazonian-vine-based tea *ayahuasca*, and that list was far from complete.

This ensemble of substances had properties and effects more diverse and complex than expressed in the terms *hallucinogen*—narrowly referring to the hallucinatory effect—or the term *psychomimetic*, producing effects that mimic psychotic symptoms. Aristotle famously said a definition should not be too broad or too narrow. As definitive and helpful terms go, hallucinogen and psychomimetic are limited, like tiny-diameter sphincters.

Instead, what they really needed was a single word embracing this group of substances. In articles and papers and from podiums (maybe make that podia) they found themselves reaching for an elusive term, and this was a most unlikely situation for a wordsmith like Aldous. Then, too, such imprecise vocabulary hampered their hunt for funds, the "Dinosaur Dough" needed from a Foundation to fund the Outsight project. Concerning the latter, Osmond and Huxley had not given up.

To land such funding one has to tell the story, and this is thorny when the hero has no name or too many. These multiple psychoactive substances were an embarrassment of riches. The publication of *Heaven and Hell* reawakened interest in the subject of mescalin etc., and excitement surged anew. Aldous decided it was high time to mint a new word.

I am taking the liberty to imagine a dialog in letters as a face-to-face incident. The following takes place eleven days after the hasty March wedding trip to Yuma. Aldous and Laura are still living at the house on North Kings Road, though their home address will soon change.

On March 30, 1956, inspired by a personal sense of let's-get-on-with-life, or perhaps motivated by a sense of podium urgency, Aldous sits down in his study to write Osmond a challenge in the form of a letter. Huxley slides a sheet of plain paper into the roller of his typewriter and, after acknowledging that soon the two men will see each other in New York (both being scheduled to attend conferences in the area), Aldous takes a few stabs at expressing his concern. Then he changes his mind, obscuring four words with the rat-a-tat of X's. (I can see them on my copy.) Aldous starts over.

He ends the opening paragraph with: "About a name for these drugs—what a problem!"[1]

What happens is a ricochet of letters back and forth across the US-Canada border—Osmond's letter dated March 24, two from Huxley to Osmond on March 30, an undated reply from Osmond around April 2—with one of Huxley's letters even chasing Osmond to a conference in New Jersey, where they would meet in mid-April.

This debate by correspondence spans several days, but I like to imagine it as if Walter Cronkite had dramatized it in *You Are There* by portraying Osmond and Huxley in a mescalin etc. naming duel, with each man championing his preferred word. I'll skip the italics but I want to note that this next scene is envisioned from the actual letters, along with speculative description and a bit of dialog thrown in.

In my mind's eye, Aldous and Humphry are together inside the thick-walled Spanish Colonial Revival home in what is now West Hollywood. It is a cool day in spring of 1956. Similar to the day in May three years earlier, when Osmond supervised Huxley's first mescalin experience, they sit on the same wicker furniture with the legs that once captivated Aldous's attention for their astonishing tubularity. The bookshelves lining the wall are the same ones that signaled a change in Huxley's chemistry when the books began to glow.

But on this day Aldous is fully alert to the conceptual rather than the perceptual matter at hand. Next to the two men sits a folding table, the kind that usually doesn't belong here because it represents a hazard for the visually impaired Aldous. The temporary table holds paper, pens, and the Liddell and Scott Greek-English lexicon, their contents familiar to the man of letters and his friend the physician. Huxley read Latin and Greek at Oxford, and Osmond is familiar with the etymology of medicine and natural science and is quite well schooled in the humanities.

Each man has prepared a list of more or less suitable terms, including those they particularly favor. They exchange their lists and scan the words. Aldous, magnifier in hand, might begin with a congenial volley:

"Let's eliminate some of these."

"Agreed. I say toss out *phantastica*. This is not about imagination."

"Very good. It's off the list." Their pens, or perhaps pencils, scratch it out.

"Jettison hallucinogen. Too narrow in scope."

"Out goes psychomimetic for the same reason. Our new word should not suggest psychosis."

They sit in silence for a while. Around this time, Laura enters the room and extends her hand in greeting to Osmond, who misses Maria but is glad for his friend's newfound happiness. She offers wine or tea and they thank her but decline.

"Thank you, but no. We must focus."

"No time to be too relaxed, nor too stimulated."

They return their attention to the now-shorter list. Osmond prefers a word he has coined for the occasion, *psychedelic*, but this is not a eureka moment. He waits for Huxley's reaction, thinking it unnecessary to point out that this is formed of *psyche* and *delos*, the latter meaning "to reveal." He expects Aldous to respond favorably to the clever idea of "mind revealing," thinking surely he will get it.

Except when looking through his magnifier Huxley misreads the word and thinks Osmond's proposed term is *psychodetic*. Disagreeing, Aldous says, "No, no, no, *psychodetic* won't do. That would be analogous to *geodetic*, which means earth dividing and so it would denote 'mind dividing.' We should convey precisely the opposite idea, that of a unifying experience."

Osmond is about to correct Huxley's error when Aldous proposes something entirely different. He points to a word on his list and proposes the verb *phaneroein*, meaning "to make visible or manifest." Aldous says, "Now *that* would be useful, because the adjective would be *phaneros* or 'manifest, open to sight, evident.'" Naturally, Aldous likes the idea of sight.

Before Osmond can pitch a rebuttal, Aldous turns it into a noun: "What about *phanerothymes*? *Thumos* means soul and is equivalent to the Latin *animus*, and phanerothyme has a musical feel." Aldous pushes his point with a ditty:

> To make the trivial world sublime,
> Take half a gram of phanerothyme.[2]

The tenacious Osmond, who is not about to give in, says, "You misunderstood me, Aldous. My word is not psychodetic but psychedelic, with the letter L—from delos, 'to reveal.'" Humphry defends his choice with a counter-rhyme:

> To plumb the depths or soar angelic,
> Just take a pinch of psychedelic.[3]

As we now know, Humphry Osmond prevailed. Psychedelic came to stand for the palette of substances Huxley and Osmond were writing about and lecturing about and experimenting with and the substances they had been seeking the funding to cautiously spread the word about ever since 1953.

Within ten years, this word would become almost synonymous with the decade of the sixties. It would rainbow across the world of visual arts and fashion and music. Yet after its birthday in March of 1956, it would be several more months before this word began to enter the mainstream. That would happen only after "psychedelic"—"mind manifesting"—made its public debut at the New York Academy of Sciences conference in 1957. Once set free, this word grew wings.

CHAPTER 16

CONVERGENCE IN CAMBRIDGE

O smond had no idea how the new decade would turn out when he wrote this New Year's note to his friend on January 2, 1960:

> My dear Aldous:
>
> All good wishes for the sixties, that sibilant S allows all sorts of pre-fixes for the questing journalist: from the *Times* to *Confidential*, each can choose an appropriate set of adjectives.

On the cusp of 1960, the term psychedelic was far from a household word. The sixties had not yet become "The Sixties," much less the Psychedelic Sixties.

At the time, Aldous was writing monthly essays for *Esquire*. Apart from his deadline for *Esquire*, among the practical items on Huxley's mind was an offer from the University of Texas to purchase his man-uscripts and first editions. These treasures were stored piecemeal in the Huxleys' home in the Hollywood Hills. Aldous had asked his son, Matthew, what he thought about the university inquiry, noting that maybe it would be a good idea to sell the library now rather than some day after he was gone, setting aside the proceeds for the Huxley grandchildren, Mark and Tessa.

Then along came a more pressing and distracting matter. In May of 1960, Aldous was diagnosed with cancer of the mouth, though after radiation treatments it seemed he had recovered and his doctor pronounced him a model patient. Recovered or not, Aldous was in no hurry to decide about selling his library. He would have to do an inventory, then obtain an evaluation from his friend the antiquarian book dealer Jake Zeitlin, though Aldous did manage to write one letter to Jake on goldenrod paper in which he asked what price he might expect for an original manuscript of, say, *Point Counter Point*—that is, if he could even find this first edition and others, along with his archives. These had been stored in various places around the house ever since the Huxleys moved from North Kings Road to Deronda Drive. After this letter, Aldous gave his library little thought.

Five years had passed since the last gathering of the Tuesday salon on North Kings Road, but Huxley's friends with "similar interests" carried on with their work, in some cases from thousands of miles apart. Then three converged in one locale.

In the fall of 1960, a celebrated author, a psychiatrist, a professor of religion, and a professor of psychology crossed paths in the university corridor near Boston. As a result, a meeting would take place between three of them on November 8, 1960, the day John Fitzgerald Kennedy was elected president.[1] What led to that meeting, though accounts vary, transpired like this:

Professor Huston Smith, who by now had been teaching at MIT for several years, arranged for his old friend Aldous to give a series of lectures in the fall semester of 1960. Their friendship had begun in the mid-1940s, when the young professor spent a day with the Huxleys at their desert home in Llano.[2]

Huxley's lecture series for MIT would be called "What A Piece of Work Is Man," its title from Shakespeare, as was often the case with Huxley's work—this one borrowed from Hamlet. Between the end of September and the end of November he would give a weekly seminar, a weekly tea party, and a public lecture (to overflow crowds). Aldous arrived in Cambridge the week of September 23.

Now Humphry Osmond entered the picture. Aldous had invited him to visit during his residency in Cambridge and offered him the folding bed in his spare room. Osmond arrived November 6. Huxley and Osmond always had a great deal to discuss, lately concerning experiments combining hypnosis, ESP, and psychedelics. ("We are also making a study of your favorite, ESP and psychedelics," Osmond wrote him on June 18.) While they were together, Aldous read him a chapter from the draft of *Island*, a section that delved into a utopian viewpoint on the matter of dying. A couple of days in proximity also meant a chance to talk about the dormant Outsight project, though someone was about to trip up their old best pony.

The man who brought that about had left a faculty position at UC Berkeley the previous year for a lecturer appointment in Harvard's Center for Research in Personality, which, like other psychology departments of the time, was influenced by the paradigm of behaviorism. En route to his new faculty position, Timothy Leary detoured to Cuernavaca, Mexico, where he ate the flesh of a powerfully psychoactive mushroom, an experience that would inform the Harvard Psilocybin Project he and fellow professor Richard Alpert launched that same year. In the fall of 1960 Leary learned that Aldous Huxley was staying down the road at MIT, and naturally he was eager to meet the author of *The Doors of Perception*, the literary godfather of psychedelics.

A curious synchronicity, coupled with a case of mistaken identity, was about to occur. First, there was a reminder that the dormant Outsight project had been blocked for half a dozen years by the gatekeepers Aldous referred to as dinosaurs and stuffy academics.

According to Osmond's book *Understanding Understanding*, this took place on election day, November 8, when he, Huxley, and Leary met. Aldous gave a public lecture that night as well. When Leary showed up at their meeting, an academic wearing a grey flannel suit with hair cropped close, he impressed Huxley and Osmond as a nice but "stuffy" Harvard man and perhaps another dinosaur, although a young one. Yet Leary's seeming decorum lent credence to his agreement with Huxley and Osmond that these compounds should be studied in a responsible manner.[3]

Osmond briefly mentioned meeting Leary in a follow-up letter to Aldous, though most of the letter was devoted to other matters. Apparently he assumed that the younger man would not risk the gains he and others had accomplished in the field. Osmond noted that Aldous spoke of "doing good stealthily," which could be interpreted as skirting authority or as not drawing attention to themselves.

As it turned out, Leary would observe standard protocols for the next two years. Even if Osmond had seen the future and said *please do not set back the progress we have made in this field* that would not have stopped the baby boom wave from responding to Leary's eventual "Turn on, tune in, drop out" mantra. Osmond had previously told Aldous how credibility, if not legality, was at stake; in a 1954 letter, at the dawn of their experiments with mescalin etc., Osmond told Aldous, "To me, the real dangers . . . are the dangers of success, not failure."[4]

Before the Psilocybin Project fell apart, before Leary's bending of policy coincided with changes in national and state laws, making it even easier for Harvard to find reasons or excuses to oust him from academia, a number of what are now considered legacy studies transpired under Leary's direction.

These include the Concord Prison Project from 1961 to 1963, which involved psilocybin-assisted group psychotherapy, aimed at reducing recidivism rates.[5] Another was the Marsh Chapel Experiment (also called the Good Friday Experiment), conducted with Walter Pahnke in 1962 and designed to investigate psilocybin as a catalyst for religious experience.[6] Thirty years later, when follow-up studies looked at studies from the 1950s and 1960s, data from Leary's research would contribute to the return of psychedelic science in the twenty-first century.[7]

Figures 3–5. Aldous Huxley while a visiting professor at MIT, November 1960. Courtesy of the family of Humphry Osmond.

Though these experiments involved the synthesized version of psilocybin, they could be traced back to the mushroom Leary had ingested in Mexico three years before and further back to the shamans or *curanderos* who, before outside visitors arrived, would have used similar mushrooms in a ritual setting with candles and smoke.

The same year as the Good Friday Experiment, the chemical god-father of LSD, Albert Hofmann, would make a journey to Mexico with ethnomycologist Gordon Wasson, whose May 13, 1957, illustrated article in *Life* magazine, "Seeking the Magic Mushroom," brought the strange fungi to the attention of the public. As Hofmann noted in his memoir, *LSD: My Problem Child*, during this 1962 journey he and Wasson experienced the magic mushrooms again, and while there they also sampled a psychoactive plant called *Salvia divinorum*.[8] The term psychedelic was proving more elastic than ever. Vision-inducing plants and fungi were still out there, waiting to be discovered.

CHAPTER 17

PURIFYING FIRE

Aldous returned from Boston at the end of November 1960, and within three months it looked as if the long-anticipated Dinosaur Dough might be within reach. Osmond told Aldous he had received word of a coalition of foundations expressing interest in funding a project based on the ideas in *The Doors of Perception* and *Heaven and Hell*.[1] They wanted to fly Huxley to New York to talk with representatives of their coalition, which included the Rockefeller Brothers Fund, the Ford Foundation, and others. It would turn out to be, if not too little, then too late.

Around the same time, rare book dealer Jake Zeitlin again reminded Aldous to look around the house and find his original manuscripts and first editions so he could do an evaluation for the pending University of Texas Library purchase. But the Huxleys were so busy traveling and lecturing that spring that they didn't follow Zeitlin's advice. Plus, Aldous was focused on finishing his latest novel, which was almost complete, apart from revision of the early chapters.

Around 7:00 p.m. on May 12, 1961, Aldous was at their home on the northern edge of Beachwood Canyon in the Hollywood Hills, and Laura was down the street feeding her friend Virginia's cat. Somewhere in the dense brush a spark, maybe from someone's tossed cigarette, ignited one dry leaf, and flames quickly spread. Fanned out of control by winds reaching 67 mph, the blaze raced up one connected canyon and down another; it surged up an incline to where the Huxley home sat on the rim, right up to their windows, a fiery tongue lapping toward curtains. In too great a panic to think about first editions and letters and papers stored god-knows-where inside the house, Aldous grabbed his manuscript of *Island*, Laura grabbed her precious violin, they each grabbed a few items of clothing, then they ran out the door and a neighbor drove them down the hill to safety.

By dawn, seventeen properties had been badly damaged or destroyed and the Huxleys' home was a charred ruin, its contents reduced to ashes. Aldous lost a lifetime of notebooks, manuscripts, irreplaceable letters, working references, and overall four thousand books, including many first editions and others with his annotations and markings. Among the lost were some of D. H. Lawrence's manuscripts. The fire had consumed all letters from others written to Aldous and his letters to Maria, dating back to their courtship days at Garsington. As he wrote to Eileen Garrett, "there is no more tangible link with the past."[2]

In a grim coincidence, Aldous had written an article in *House and Garden* in November of 1947 called "If My Library Burned Tonight," with a list of books he then felt he would need to replenish. It was doubtful, though, if he would replace them at this point in his life. As a temporary measure, Huxley moved in with Gerald Heard in Santa Monica in order to finish the novel. A publication date hung over him despite this personal loss, yet plunging into editing and getting back inside his own head probably provided solace. For her part, Laura moved in with Virginia, whose home had also been destroyed. They set up housekeeping in a rented place that would soon become a home for their merged families.

In coming months, the Huxleys would talk about rebuilding on the burned-out Deronda Drive lot, but neither had the heart for such a project then and they never would. The default was Virginia's rented two-story 1930s-era Mission-style residence on Mulholland Highway. In a way rootless, in a way strangely free, as if starting to untie attachments to this world, Aldous began traveling again. One trip would be to Eileen Garrett's International Conference on Parapsychology at Le Piol in the summer of 1961.

It is February of 2007 and I am in John Smythies's office at UC San Diego, picking his brain. He laughs softly when he recalls how Huxley's forthcoming novel *Island* was the talk of Garrett's parapsychology conference that summer of 1961. Though the official publication date was a few months away, apparently the word spread that their hostess was the model for the outsized oracular character Huxley called the Rani.

Eileen Garrett was the second of two influential women who played major roles in Huxley's novels, early and late in his writing career. Lady Ottoline Morrell left her mark on his first novel, *Crome Yellow.* Eileen Garrett is a presence in *Island,* his last. In the country house setting of *Crome,* Huxley portrays members of an elite class in a country estate facing change after the Great War. In the isolated setting of *Island,* change has landed on the distant shores of the fictional nation of Pala. An advanced society founded on a balance of body and mind, Pala's culture of lifelong learning blends liberal studies, physical conditioning, hypnosis, and a gentle regimen of psychedelics. But this is a utopia whose resources are coveted by outsiders. Such bounty harbors the seeds of its downfall.

When Aldous attended Eileen's parapsychology symposium at Le Piol in the Provençal sunshine in the summer of 1961, so soon after the fire, I think his visit closed a circle. It suggested a reprise of Garsington Manor, where Aldous had met Maria, and even had a hint of retreat, as in the Trabuco monastery years. Le Piol was a magnet and gathering place where people met and talked and shared innovative and sometimes radical views.

"The site was perfect, the food and wine delectable, the company was stimulating," Smythies, who was there that year, recalls.[3]

Decades earlier, the flamboyant Lady Ottoline had considered

Aldous's portrayal of her a betrayal. John tells me that when Aldous made the Eileen Garrett character a flamboyant spiritual leader Eileen didn't mind. Looking at photographs of the 1961 conference, I see one of Eileen with a furrowed brow, deep in thought.

Aldous had begun what became *Island* in 1956, shortly after the "mescalin etc." word duel with Osmond; it was published in March of 1962. Devising the useful term psychedelic might have helped Aldous get moving on a book he had long intended to write—not the dystopia of *Brave New World* with its government-issued drug soma but a utopia with *moksha*-medicine, named for the Indian philosophical term for liberation. One ingests this in an initiation, revisits the experience perhaps once a year or so to gain insights throughout life, and ultimately takes the medicine in closure.

Some of Huxley's ideas about unorthodox healing show up in *Island.* He was working on this book in the late 1950s when he told the team of journalists from UCLA, "One of my principal characters is, like Darwin and my grandfather, a young scientist on one of those scientific expeditions—a Scotch doctor, who rather resembles James Esdaile, the man who introduced hypnosis into medicine."[4]

Island is in part a cautionary tale about trusting a utopian vision. The noble Pala experiment is bound to fail, as did the utopian commune at Llano del Rio, but Huxley's novel remains a prescription for life worth living, a celebration of human potential in a world confronted by change.

Now I am looking at another photograph taken at Le Piol that summer. Aldous sits on the hilltop terrace midinterview, a microphone positioned on the table. I see what may be lavender fields in the background. It has only been a few weeks since his home and precious manuscripts and letters were destroyed and Aldous looks weary.

His utopian novel *Island* officially reached the bookstores in March of 1962, and that same month Aldous gave lectures at, among other places, the University of Alabama at Birmingham. This had been Timothy Leary's alma mater. It was where John Smythies's career in neurology would take him in 1972 and where Osmond would end his career as director of the Alabama State Psychiatric Hospital. That spring Aldous would undertake other trips and lectures, including a talk to scientists at Los Alamos. I think it is reasonable to say that whether he was talking about Shakespeare or technology or ecology or the threat of overpopulation (before the Pill), his Visionary Experience talk was a centerpiece of his program.

In June the cancer returned. Aldous told Osmond but asked that it be kept quiet. He underwent an operation to remove the cancer, followed by cobalt treatment, a cycle that would repeat at the same time the following year, as if his spring lectures worsened his condition, yet somehow he carried on both times after surgery and treatment, though the second time he made a poor recovery.

Somewhat remarkably, in a letter to his brother on June 2, 1963, Aldous said he had begun "feeling my way into a kind of reflective novel," buoyed by a mistaken sense of rebound and thinking he would have the luxury of years in which to write. The thought may have enlivened and cheered him, and maybe, under those circumstances, as a writer that's what I would want to do—look forward to a project. In reality, his health would restrict him to composing shorter pieces. In 1963 he nonetheless managed to log many miles of travel early in the year, and after the second surgery that summer he and Laura flew to Stockholm for a meeting of the World Academy of Arts and Sciences.

Dr. Humphry Osmond was with him there. Now I am looking at one of the late photographs of Aldous, this one taken by Osmond in a hotel room in New York in March of 1963. It shows an almost ethereal Aldous, his eyes close-focused on the page as he reads by a window in ambient light.

Following Stockholm, Aldous visited his brother Julian in London and retraced familiar steps from his childhood. He flew to Italy and met Laura there, and afterward they returned to Los Angeles at the end of August. After this he would stay close to home and work on an essay called "Shakespeare and Religion."

Though the clock was running down, there was still time for Huxleyan humor. Gerald Heard recalled the day he brought Aldous a copy of his latest book, which he had personally inscribed. The exchange below alludes to what King George III said when historian Edward Gibbon, author of *The History of the Decline and Fall of the Roman Empire*, presented the king with a copy.

As Gerald handed a copy of his own new book to Aldous, he said, "What! Another big, fat book?"

Aldous fired back, "Another damned, thick *square* book."[5]

On another day, Gerald made a visit to the house at the top of Beachwood Canyon to see his old friend, whose cancer by the autumn of 1963 had reached an advanced stage. Heard described Aldous coming down the stairs from his bedroom study, still elegant and agile. He tired easily, Heard said, but his interest in life had not waned.[6] The two companions had crossed the Atlantic together in support of the Peace Pledge Union, they had crafted the mission for Trabuco College, and that day they walked together on Mulholland Highway, though not straying far from the house on the canyon rim.

On November 22, 1963, I was wearing headphones in a language lab in Los Angeles when someone entered the classroom and whispered to the instructor. Looking shocked and pale, he turned off the master switch and said, "I need your attention. President Kennedy has been shot."

A few miles away from where I sat, Huxley's life also was slipping away, in the home under the Hollywood sign. In one room the television was dialed low to prevent Aldous from hearing that in Dallas, three hours earlier, an assassin's bullets had taken down the president.

Figure 6. Aldous Huxley and his brother Sir Julian Huxley, 1962 at Onet Cottage. Courtesy of the family of Humphry Osmond.

Figure 7. New York, 1963. Huxley writes by focusing his eyes just inches from the page. Courtesy of the family of Humphry Osmond.

Around midday, and despite the doctor's initial reluctance, Laura gave Aldous what he had asked of her in a scrawled, hand-written note. Laura prepared and injected the intramuscular dose of LSD. Aldous had all but scripted this scene in *Island*. With this last rite, fiction became the amber of biography.

Following Huxley's passing, his family arranged for his ashes to be interred in his parents' grave at Compton, Surrey. Later on, Maria's ashes would be returned to England to rest alongside his. A memorial was held in London on December 17, 1963, and remembrances were collected in a memorial volume. Among the family and friends who contributed essays were Humphry Osmond, Gerald Heard, and Christopher Isherwood. In his remembrance, Isherwood wrote about Huxley's "fearless curiosity."

Aldous questioned unceasingly, and it never occurred to him to bother about the neighbors. They laughed at him for consulting unlicensed healers and investigating psychic phenomena; and it was true that many of the healers proved to be wrong and many of the mediums frauds. That was unimportant from Aldous's point of view. For his researches also brought into his hands some very odd and precious pieces of the jigsaw puzzle of Truth; pieces that may not be officially fitted into the main pattern and recognized as scientifically respectable for many more years to come.[7]

PART III

AFTER ALDOUS

CHAPTER 18

BACKPEDAL ON LSD

Not very far from the Hollywood Hills, where the Huxleys lived together until Aldous's death in 1963, two Southern California psychiatrists whose names have come down in the annals of psychedelic lore conducted their clinical research.

One was Dr. Sidney Cohen. Laura Huxley had called Cohen the night before Aldous died, knowing the end was near for her husband. She asked the psychiatrist, who was known for his research with LSD, whether he thought she should offer Aldous a dose of LSD; he told her he had given it to two people who were near the end of life, and for one it seemed to help and for the other it seemed to make no difference.[1]

Since the mid-1950s, Cohen had been affiliated with UCLA and the Veterans Administration Hospital in West Los Angeles, where he was chief of psychosomatic medicine. In the late '50s he conducted a government-funded study of the potential of LSD for treating psychotics and prisoners, though he was also interested in the drug's effects on creativity. He also would preside over LSD sessions with a number of well-known individuals, including Clare Boothe Luce, the wife of cofounder and editor-in-chief of Time, Inc., Henry R. Luce.[2]

His counterpart in Beverly Hills was Dr. Oscar Janiger, whose private LSD-assisted practice would involve an estimated nine hundred patients and numerous celebrities, among them actors Cary Grant and Jack Nicholson. Between the mid-'50s and mid-'60s, in addition to his private practice, Janiger conducted substudies on the use of LSD, such as one looking at artists and creativity when under the influence of the drug.[3]

Cohen and Janiger were among a number of mental health professionals who contributed to the growing body of knowledge about psychedelics—mainly LSD—that led to publication of more than one thousand professional papers involving tens of thousands of subjects.

In a follow-up study conducted in the 1990s, Rick Doblin, the founder and director of the Multidisciplinary Association for Psychedelic Studies (MAPS) tracked down forty-five people who had participated in Janiger's original LSD research. Recalling how he had come to participate, one former subject said that Janiger, who also taught at the College of Osteopathic Physicians and Surgeons, had invited Huxley to give a lecture there. "It was right after *The Doors of Perception* had just been published," said the respondent. "He asked me if I was interested. I came, and he was talking about his LSD experience. Afterwards, Oscar indicated that he was going to apply for a grant, and I told him I'd like to be a guinea pig."[4]

The mention of a grant, not to mention a guinea pig, raises the question of informed consent and how such studies were funded. During the early days of psychedelic experimentation in the United States funding was fluid. LSD (and later psilocybin) manufacturer Sandoz gave support, but mainly it came from the US government, through agencies such as the Veterans Administration, National Institute of Mental Health, as well as, clandestinely, in programs devised by the CIA in its notorious MK-ULTRA program.

As has been widely reported and well documented, the CIA responded to rumors that the Soviet Union was using mind-control drugs by launching its own research. The intention was to explore interrogation methods (sometimes called brainwashing) by using substances such as LSD, often combined with techniques like hypnotism. Much has been revealed since then about human guinea pigs unaware they had been given powerful hallucinogens.

Huxley had heard early rumors of one such government program when he traveled to North Carolina to deliver a series of lectures in the fall of 1954 at Duke University, the location of his friend Dr. J. B. Rhine's parapsychology lab. Huxley wrote to Osmond that, while in conversation with Rhine, he had heard about the National Institute of Mental Health experimenting with lysergic acid.[5]

Records since released to Congress show that, beginning around this time, the CIA supported psychedelic research through programs with major universities.[6] If you were a psychiatrist or university-affiliated researcher who landed a grant to study hallucinogens, you might not be aware of the wellspring behind your funding. To some working in psychedelic research, the government backing portended a bright future; others, like Sidney Cohen, saw trouble on the horizon. In 1962 Cohen would be one of the first to warn about psychedelic drugs going mainstream in an emerging black market.[7]

By that time, the once-progressive social climate was beginning to change in Canada, no longer encouraging radical mental health experimentation. By 1963 Osmond would move to the United States, working with schizophrenic patients rather than psychedelics, as director of the Bureau of Research in Neurology and Psychiatry at the New Jersey Psychiatric Institute in Princeton.

In an abbreviated timeline, you might say that the retreat of psychedelic research started the year Huxley died, 1963, the same year professor Timothy Leary was ejected from Harvard. Seven years later saw the landmark Comprehensive Drug Abuse Prevention and Control Act of 1970, which turned into a campaign then president Richard Nixon called the War On Drugs. Another way of looking at that seven-year span between the two is to recall what was happening simultaneously in drugs, laws, and social history.

To step back a bit, the Harrison Narcotics Tax Act of 1914 brought into play the requirement of prescriptions for certain opioids and also required keeping records and mandated the label requirement of "Warning: May be habit forming." Humphry Osmond was familiar with the Harrison Act when he joked with Aldous about surreptitiously importing the peyote-derivative mescalin in 1953, the first time he came to Los Angeles.[8]

Jump ahead to 1962, when new drug law amendments were enacted during the same year black market LSD started to appear on both coasts. The Kefauver-Harris Drug Amendments may have been partly prompted by concern over psychedelics, but this legislation also came soon after a drug called thalidomide was pulled from the market following the birth of thousands of deformed infants after their mothers had been given the drug to alleviate morning sickness. Among other changes, the Amendments required clinical studies by well-qualified experts and the informed consent of subjects.

As use of LSD and psilocybin expanded, some states required that only medical doctors conduct human drug research, which for psychotropic drugs meant psychiatrists, who are also physicians. That same year a group of Harvard professors objected to the way drug experiments were being conducted in the Harvard Psilocybin Project, by Timothy Leary and his colleague Richard Alpert (later called Ram Dass). At around the same time, a new Massachusetts Food and Drug Division law required that such drugs could be administered only in the presence of a physician, though Leary and Alpert both had PhDs.

Other circumstances contributed to Timothy Leary's ouster: an entrenched Behaviorist-Freudian faculty suspicious of a biochemical approach and (as the story goes) Leary and Alpert taking psilocybin with undergraduates—though the official reason for ending professor Leary's contract in Spring of 1963 was the objection that he had missed too many class lectures.[9]

Meanwhile, the winds of social change blew in from all directions. The following year saw the Freedom Summer Project, a student-led campaign to register black voters in Mississippi. In San Francisco, where demonstrations focused on unfair hiring practices, hundreds of us packed into the lobby of the luxurious Sheraton Palace Hotel in San Francisco to protest racial discrimination. On May 2, 1964, as many as one thousand students in New York and an estimated seven hundred of us in San Francisco marched in the first major student demonstrations against the war in Vietnam.[10]

With no fanfare, in 1965, the psychology department of my campus, San Francisco State, had established the Psychedelic Research Institute. Its goal was to explore the creative power of LSD.[11]

My idea of a trip, at that point in time, required tangible transportation. In the summer of 1965, my boyfriend, Marc, and I set out with student Eurail passes for a three-month adventure, not knowing that the journey would change my life. This was not a tour, and we had little in the way of resources as we stumbled through nine countries following the frugal advice of *Europe on $5 a Day.*

The relentless travel and living on the cheap wore us down. We quarreled, and by mid-summer in Italy we had parted. I had also lost my outward focus after three months of gazing at too many spires and gargoyles, and I was sleep-deprived from tossing too many nights on lumpy youth hostel beds. For a while I'd been sick and feverish, and by this time my travel journal had more blank pages than entries, as if the present had begun to fade.

I now understand how exhaustion and illness can create a vulnerable condition, when extraordinary perception breaks through a crack in one's rational armor. Some spiritual disciplines create this condition through fasting and chanting, swaying and dancing, by imposing punishing excursions into the desert, any or all giving rise to the altered states that were of such great interest to Aldous, and that is what happened to me.

Alone in Florence on a hot summer day, I boarded and found myself alone in a train compartment where someone had cracked open the window. The engine idled as I caught my breath, being winded after rushing from the youth hostel to the station.

Then something odd happened. Moments after I sank down on the padded seat, a moth flew through the window. The creature circled around the compartment and landed on my fingertip as if this was its intended destination. A large hawk moth, it was homely but beautiful, with a span almost the width of my palm and wings seeming woven from silk in shades of gold, black, and copper. Imagine a peacock in sepia.

I felt a twinge then a pulsing sensation, as it clung to my fingertip and slowly fanned its wings. As I stared at the moth's slowly fluttering display, the world around me dissolved from complexity into a unity of pure simplicity, and I was overcome by a loss of self as who I was flowed into this presence of nature. Maybe less than a minute passed by, though I can't say. At some point, other travelers entered the compartment, and I watched as the moth banked through the crack and flew away.

Afterward, I rarely shared the story of my experience with the moth because I found that the retelling made me sound slightly unhinged. That's the problem with verbalizing the wordless. In one instance, I told my tenderly personal tale to an ill-chosen listener and he mocked me. That relationship with Marc hadn't survived the summer, and by October I had a new boyfriend, CJ, who introduced me to LSD.

A year later possession of LSD became illegal in California, on October 6, 1966, and on that same day a crowd of a thousand or so gathered in Golden Gate Park to drop acid to the backdrop of Big Brother and the Holding Company, though my lasting impression is of seeing the Merry Pranksters' bus. I avoided the questionable giveaway acid because, back at our small rented house, CJ had a large bottle of the stuff. When I asked where he bought it, I found he had spent the summer of 1965 on a small live-aboard boat in Newport, by then an epicenter for California's black market acid. One day when

CJ was sitting on the deck of his boat, a dealer named Chemical Al appeared, hotfooting it through the marina to escape an officer in pursuit. Running past, he threw the bottle to CJ, saying something like, "Catch it. A gift."

If you look closely at the album cover of *Sgt. Pepper's Lonely Hearts Club Band*, quaint though an LP may seem these days, you will see Aldous Huxley in the montage of famous and infamous faces. The iconic Beatles album with the postage-stamp faces came out in June 1967. Across the city, from the Haight to the Golden Gate, this album became the sound track for the Summer of Love.

On the federal level, possession of LSD would become illegal in October of 1968, with passage of the Staggers-Dodd Bill, leading to the 1970 Comprehensive Drug Abuse Prevention and Control Act. President Richard Nixon's special message to Congress in June of 1971 concluded with the metaphor of war that launched a communications campaign, the so-called War on Drugs.[12] Though failing to vanquish the underground drug culture, in part because of new synthetics, the war on drugs succeeded in sending sanctioned psychedelic research into retreat.

One day, early on in this book project, my father and I were sitting under the grapevine when a spider spooled down on its tangent line and landed on my arm. I practically upended the chair in my rush to stand up and flick him off. Dad laughed as the brown dot scuttled across the concrete floor and said, "That kind can't hurt you."

"Sure, garden spiders benefit mankind. I know," I said crankily, feeling the adrenaline, and I explained how once I had even tried an experiment involving dozens of spiders in an attempt to overcome my fear. The setting was a spider-rich section of Golden Gate Park, the Rhododendron Dell. That day, after my friend CJ and I arrived at the Dell, I sat down cross-legged and in no time saw a solitary red spider sitting on a wood chip. Hand like a gangway, I invited the spider on board. "The truce has lasted since then," I said, "except when I'm surprised. Actually, there was more to it. Earlier that day I took LSD."

It occurred to me that Dad and I had never talked about psyche-delics, maybe because in the mid-'60s we had barely stayed in touch. Remarried by then, he was working for Litton on the design team inventing components for missile-guidance systems.

"Did you ever try LSD?" I asked.

"Not me."

This was a surprise, considering how he had urged me to experi-ment, considering how he had been a friend of Aldous around the time of Huxley's first mescalin encounter.

"I suppose there was a slight risk," I said, thinking back on how most of my friends had experienced no problems, though one boy had been hospitalized after LSD apparently triggered a psychotic episode.

"If you want to know my reason," he said, "I didn't want the wrong kind of drugs to interfere with my healing work."

In the 1970s, after years of sidestepping the experiments and dis-appointments of the 1950s, my father had taken up healing again. He began by holding meetings a few nights a month after the end of his workdays. A reporter at the Hearst-owned *Los Angeles Herald-Examiner* got wind of it, and an article about my dad ran on July 16, 1977, with the headline "No Faith Required for Cure, Healer Says." When I read it years later I saw that, at one point in the half-page interview with Howard Thrasher, the reporter had written:

> Thrasher also says that tuning into "it"—"you could call it God or Jesus or strawberries"—may not result in healing power but some other type of extra sensory perception.

Five years after this article appeared, I would, for the first time, see my father practicing his method of healing. I had moved east to take a job in the New York publishing industry, and two years later I flew home one weekend to attend a reunion. I stayed with my mother in my grandparents' old bungalow in Long Beach, the one with the patio and big garage, but I wanted to see Dad before catching a red-eye flight back to my other life in Manhattan.

He and I agreed to get together, and over the phone he gave me his address in the San Fernando Valley, not far from where he worked at Litton. After I located the place, I rang the bell. "It's unlocked," he called from inside the rented house where he lived alone, at the time separated from my stepmother. I walked into the living room and found it empty except for two folding chairs and a TV tray with a thick, black three-ring binder on top but no TV set in sight. He appeared from the back of the house and embraced me with one of his taut, wiry hugs, then asked, "Well, how was the class reunion?" I was surprised that he remembered, though my mother had asked about former classmates whose names and faces she still recalled.

Dad saw me looking at the black binder on the TV tray. There wasn't much else in the room to look at. He said, "I've written this down over the years, trying to figure out how this damned healing process works."

I saw a title on the side, written in his precise draftsman's letters: "The Art of Perceptive Healing." Recognizing a chance for some rare father-daughter bonding, I plunged ahead. "Dad, would you like me to edit your book?"

He looked puzzled. "I don't care if this is published."

It was a fat binder and not just a few pages of jottings, so I had to ask, "Why else would you go to so much trouble?"

"To give you this notebook, or maybe you could call it a manual. But you would need to learn by watching, and I'd need at least week of your time."

I was stunned and touched, yet at the same time alarmed. Apparently he'd had this in mind all along, to pass this legacy on to me, his only child. But taking a manuscript back to edit was one thing, versus a detour in my too-busy city life. That prospect, the detour, made me want to flee. Of course, within hours I'd actually be gone and soon on the ground three thousand miles away.

"Think it over," he said. "I have a session tonight and I'm expecting about a dozen people. Why don't you come and at least see how this works."

I mumbled something about my red-eye flight being at ten thirty, begging off, but he said, "I think you'll have time." The meeting would start at seven, he said, so if I stayed a short while I could still make it to LAX with enough time to drop off my rental car. I weighed what he was asking against saying no and thought, *Well, sure, I can do this much.*

"No need for formalities," he said. "When you need to leave, walk out."

Shortly before seven I parked beside a small office building located in Calabasas in the West Valley and entered a standard 1970s-style office suite, with carpet the color of gruel. Most of the dozen or so seats were taken. I took one near the door. A middle-aged man sat down beside me and asked, "Have you been here before?"

"No," I replied in a barely audible voice, hoping to remain anonymous.

Talkative himself, he volunteered that he had no current medical complaint but had attended several of Howard Thrasher's sessions after reading the *Herald-Examiner* article. "He is most unusual," the man said. "I've seen him help several people, and I'm trying to figure out how he does it." Then he asked, "Why are you here?" but I had an excuse for silence because just then the session began.

From the front of the room my father smiled at me, and, after a brief greeting to the assembled group, he began. He approached his first subject, a middle-aged woman, and began a kind of rhythmic breathing, his arms and hands moving in a slow circular motion, hovering near her without touching, as if tracing an invisible outer skin. I was relieved to see that this was a subtle process, with no Elmer Gantry–like fire-and-brimstone antics or anyone crying out. He had told me it was usually gradual, though at times results came more quickly. I watched for a while as he went through the same movements with another subject, a man, then I checked my watch and slipped away.

While driving toward the airport my inner skeptic and inner believer faced off. I know skeptic doesn't mean denier but one who doubts, and I know there were/are/will be plenty of quacks and frauds, but that night my father didn't seem to be one of them. He personally paid for the office space where the sessions took place. He did not accept payment. His aerospace job paid well. When it came to media exposure, the press found him. He didn't seek fame, and he didn't even care about publishing the pages in the binder, for chrissake, so why would he go to such bother, my inner believer asked, unless to him this was real?

I arrived at LAX and dropped off my rental car with ample time before the call for boarding. A few hours later the cabin lights blinked on, and a cart loaded with orange juice and coffee rattled down the aisle. Soon we were on the approach to JFK.

CHAPTER 19

WITH LAURA UNDER THE HOLLYWOOD SIGN

I showed up frazzled one day and in a rush of affection wrapped my arm around my father's shoulder. He felt fragile yet resilient, this man in his mideighties who still did strenuous daily floor exercises.

"Take a look at this," I said, hauling a book from my handbag. It was Laura Huxley's memoir, *This Timeless Moment*, a recollection of the Huxleys' marriage from 1956 to 1963. I opened to a page and there was the image of Huxley's hand with the caption, "Aldous wrote his name on this photograph of one of his hands shortly before his death."

Dad retrieved a pair of reading glasses from the pocket of his retired-aerospace-guy short-sleeved shirt and peered at the page. "This is not one of mine."

"I didn't think so."

"Wrong technique." He pointed with a blunt fingertip. "He used a direct flash—see the flare?—and overexposed the palm. Not much there."

Finding this photograph in the book meant that even if she was a late arrival in my story focusing on the Tuesday night circle of the early 1950s, I definitely needed to have a word with Laura Huxley. But first I had to find a contact number and arrange for an interview—or, given her emeritus status in LA, you might call it an audience. I tried various means, such as going through her publisher, then going through some of my book industry colleagues, and I had just about given up when by a fluke of the calendar I finally found a way in.

It came through Don Bachardy, the artist and longtime partner of Christopher Isherwood, whose 1962 drawing of Aldous had appeared in the 1998 edition of the posthumously published *Jacob's Hands.* This connecting dot took the form of a newspaper announcement saying Bachardy would be the featured artist at the Grand Central Gallery in Santa Ana.

After I introduced myself and briefly told him about my Huxley's Hands project, he offered to put me in touch with the Huxleys' longtime literary agent, Dorris Halsey. When Mrs. Halsey called, I told her about my project and the link between hands and my dad and Aldous.

"Interesting," she said. "Mrs. Huxley has a framed picture on her wall."

"A picture?"

"A photograph of Aldous Huxley's hand."

Within a week or so my hard-won interview delayed by so many blind alleys had finally been arranged. Soon the calendar rolled around to the designated day in October of 2001, and Laura was expecting me.

I walked up the steep driveway of the 1930s-era home on Mulholland Highway and looked up, realizing that I was retracing Aldous Huxley's footsteps. In a study-bedroom on the second floor he had written one of his last letters, and it happened to have been addressed to Humphry Osmond. In that upstairs room he had composed his final essay, "Shakespeare and Religion."

When Laura greeted me at the front door, I was aware of her soft Italian accent. She was a petite woman in her eighties, about my father's age, with a neat white halo of hair, wearing slacks and blouse in a serene palette of aqua blue. Once inside the foyer I was struck by the flood of sunlight pouring through tall windows onto upholstered chairs and sofas in the living room, gilding a rocking horse and potted orchids, bathing the pale walls in a lemony light, beaming onto white carpet and walls. This was the bright interior effect she had first created for her visually impaired husband.

The next thing I noticed was the framed black-and-white photograph on the living room wall. You couldn't miss it. Just seeing the photograph made me feel it was worth the trip, but another surprise was in store.

I had brought along a copy of the first photograph, the one my father made when Aldous was in his midfifties, and by all accounts perhaps at his healthiest. The image on the wall looked more emaciated. It had been taken, Laura told me, at her husband's request but in the late stage of his final illness. We looked at mine, then the version on the wall, and clearly one looked more robust than the other.

Then, too, the photographic print I had with me was more analytical, a blow-up, a reverse negative in white on black. My father's measurements and his grids were annotated along the sides, so Dad's version seemed to me less poignant, less human somehow, than the later photograph displayed on the wall, the one taken when chances are he knew he was dying.

After we were seated on the sofa, she asked about my project and my reason for being there, so I stated the obvious. "I know Mr. Huxley took an interest in the hand," I said. When I told her about the extent of my father's collection, she responded with a slightly idiosyncratic choice of words: "You are making a research project and someone might be interested. You have almost a thousand hand photographs? That would not be too difficult to scan into the computer."

I had been wondering if some hidden value might remain in these images. Maybe someone using today's technology could benefit from revisiting the hands of the schizophrenic patients at Norwalk and Pomona hospitals; after all, the photographs are sharp and clear, with privacy preserved, identified not by name but by number, and the quest to understand schizophrenia continues to this day.

I mentioned how my father had been present at many of the Tuesday night gatherings in the early 1950s. It seems Laura had attended one or two of these, she said, and she thought she might have met my father when they were both in their thirties.

She noticed at one point while we were talking that the sunlight had shifted. "Do you want to change your seat?" she asked.

"I think so. The sun is straight in my eyes."

"Then we will go outside." She led me to the terra cotta patio and we chatted in the shade, my shoulder nearly grazing hers because she spoke so softly I had to lean in to understand her words. Above us, across the canyon, loomed the white three-dimensional Hollywood sign, the elephantine geometry of its H uncanny in that it was both familiar and disturbing, for you usually see it from a distance of miles and here it was magnified.

After a while we returned to the living room, where the strong beam of sunlight had become more diffused. We continued our conversation for a few more minutes, though I didn't want to tire her. I asked a few questions about incidents in Huxley's life in the early 1950s, but much of this was before their marriage, and we talked about her books and her work—Laura had become a psychotherapist, and her 1963 self-help book *You Are Not the Target* had enjoyed considerable success. She seemed to enjoy our conversation and so did I, but I was still trying to fathom the meaning of the hand photograph on the wall.

After I thanked her and rose to go, she said she couldn't picture my father's face, although maybe their paths had crossed one Tuesday night on North Kings Road. I had not thought to bring along a photo of him taken at that time; still, they were near contemporaries. Laura walked me to the door, and, as I headed down the driveway with an invitation to come back, she called out to me, "Say hello to your father."

If my father and Laura had met back then, I would have been about nine years old, and I wasn't the only child whose hands had been coated with a minty white solution, nor the only one to hear discussions way over the head of a nine- or ten-year-old. Eventually I would connect, or reconnect, with two others who had been children with me back in the '50s.

This came up in an indirect way one day when Dad and I were hanging out on the patio and my mother joined us. I asked her to help me solve a memory puzzle. "I remember one time we drove to a large house somewhere in Hollywood," I said, fishing for fragments, and almost any house would have seemed large compared to our cottage. "I recall falling asleep in what must have been a guest room."

Since this image had surfaced I'd been indulging in a tantalizing possibility. Did I sleep in Aldous Huxley's guest room? This would be the same room where Humphry Osmond had stayed during the mescalin-initiation week of May 1953, the trip that prompted Aldous to write *The Doors of Perception.*

"Were we at the Huxleys' home on North Kings Road?" I looked around to see how my question registered. Dad's forehead took on his perplexed look. My mother smiled because she had the answer.

"That was a beautiful home, but it wasn't the Huxleys'," she said. "Aldous and Maria rarely welcomed children on Tuesday nights. No, you're remembering the home of Ed and Barbara Muhl. They had small children and didn't mind if we brought you along."

"Ed was the studio chief of Universal Pictures," Dad added. Muhl had been the head of the studio's production operation for twenty years, overseeing such films as *To Kill a Mockingbird* in 1962.

"Do you remember the time we visited the Muhls' other place, the ranch in Saugus?" She turned to Dad for confirmation. "You and their son Lee ran all around the ranch. That was the day you cut your leg on a roll of barbed wire."

"I *do* remember that!" I said, flashing on an image of myself playing chase with a boy my own age and not seeing the hidden obstacle that caused so much pain. This was another of those oddly time-warpish moments, sitting with my parents fifty years after the family split up and talking about a shared incident. I showed them the scar on my leg, and Dad looked newly alert. It struck me how even a distant father remembers with a jolt a time when his child was injured, but this was fleeting. He quickly cut to the topic of experiments.

"One night, as I recall, Ed Muhl arranged for a crew to film Sophia," he said, referring to the woman who was Aldous and Maria's primary trance medium. "The film crew used black-light photography," though I would later learn that the film crew was probably arranged through some other Huxley acquaintance in the film business.

I didn't think much more about this crazy quilt of family anecdotes until months later when I tried to track down the Muhls. From online research I found that Ed Muhl was no longer alive, nor was his wife. I ran through Dad's index of names looking for a cluster, and there I found Barbara and Ed, hand photographs 397 and 398, then Leigh Muhl, 399, followed by the entry for Maria Huxley's youngest sister, Rose. But wait—I happened to glance a page farther in the notebook and there was an entry for a Lee Muhl, number 459. So which one was my former playmate that day when the barbed wire snagged me? Was it Leigh or Lee? Who was this other person?

First I searched Google for Lee Muhl and found out that he worked in the film industry, like his father before him. Then I found him on LinkedIn and sent a message: "We were childhood friends." Sure enough, he got back to me and I told him about the one thousand hands and the Tuesday night experimental salon our parents had attended at the house on North Kings Road.

"Great story," he said. "I can help with some of it."

What about those names?

"That's actually me," he said. "The birth spelling of my name is Leigh, and it was only when I was eight or nine that I changed it to Lee due to a teacher's role call on day one of a fourth-grade class.

She pronounced my name as 'Lay Mule,' so I went home and told my mom 'the hell with this Leigh stuff.'"

That explained the two names. Lee did not recall having his hands photographed, much less twice. We were too young, I suppose, to be meaningfully included in my dad's study, much less grasp what he was getting at.

But there was a kid who did get it, and his name was Siggy Wessberg. I'd say maybe he got it because he was a year or so older than Lee and I were at the time. Siggy was the Huxleys' nephew, the son of Maria's sister, Rose, and had grown up in Llano near the Huxley ranch. Aldous and Maria bought a house for his mother nearby and it, too, faced the Sleeping Elephant in one direction, while in the other direction the San Gabriel Mountains stood sentinel over a sere landscape of creosote bushes and Joshua trees and spectacular sunrises and sunsets.

Siggy not only knew about his Uncle Aldous's interest in hands but he recalled hearing the grownups talking about it. Apparently this hand business made an impression, because I'd read a mention of it in the transcript of David King Dunaway's interview with Siggy, archived in the Huntington Library in San Marino, and now I wanted to talk to him myself.

It took me a year to find Siggy. I found him through the current owner of the Huxley ranch, who, I learned on the day I had a chance to roam the house and grounds, was a Huxley buff herself. Better yet, she happened to have Siggy's phone number, so for me that was a doubly lucky day. When I called Siggy, he confirmed the story I'd read in the archive.

The occasion, he recalled, took place in a Polynesian-themed restaurant in Hollywood. Chances are it was Trader Vic's—the upscale eatery that claimed to have invented the Mai Tai circa 1944 (a claim also made by its primary competitor at the time, Don the Beachcomber). The mid-1950s marked the peak of the Tiki-culture fad, a time when one of my uncles even converted part of his garage into a blowfish-lit, rattan-chaired tropical lounge he and my aunt used for entertaining.

Meanwhile, the boy named Siggy sat among the grownups at a Tiki-themed restaurant at a dinner hosted by his Uncle Aldous, circa 1954. Also present was his mother, Rose, and a few others, including a special dinner guest. Siggy said he remembers that Aldous and this other man were both engaged in the study of hands, and chances are the man he remembers was my father.

Siggy's other impression of that evening was the way several waiters gathered around the table when Aldous spoke. A small crowd typically formed like this because Aldous had a way of explaining so that even if the subject was obscure, almost anyone could understand. That night, Siggy recalls, Aldous offered his take on the hand study: It takes observation of at least twenty to thirty hands before you can determine an average form, he said. Once an average form is established, you will begin to see the differences.

Not until decades later would my father began to explain those differences to me and explain how they could be quantified and demonstrated statistically in graphs and curves. Despite his own personal encounters with the uncanny, and there had been several, my father had carefully designed the hand project so there would be not a whiff of the occult about it; he used photographs and measurements and statistics to determine whether there was a correspondence between human traits and structural patterns in the hand. He found them, and so had Professor Gengerelli in his replication study.

Dad and I were sitting on the patio when he brought up the ill-fated conference at the Miramar, that turning point when, after he and Professor Gengerelli had presented their findings, the audience had all but booed them out of the room and Gengerelli had shelved the controversial study rather than risk his career at UCLA.

I knew that afterward my father had stored away the photographs and charts; he'd vowed to steer clear of any experiments on the fringe of science. He held to this for about a decade, focusing on his day job. Subsequently, he was credited with nine patents for Litton Industries, designs contributing to the platform for gimbal technology, which is to say components for missile guidance systems. At Litton in Woodland Hills no one knew that he often solved mechanical engineering problems in dreams.

Sitting with Dad one day on the patio, I learned that Dr. Gengerelli finally published an article based on their abandoned study almost two decades later, after the professor retired. It appeared in 1979 in the *Journal of Psychology* with the co-authored credit of J. A. Gengerelli and H. Thrasher, and the title "Termination Locus of Palmar Main Line of the Left Hand in Relation to Mental Pathology." To me, it had been the little triangle that seemed to indicate schizophrenia. In the journal article, Gengerelli summarized their findings from a quarter of a century earlier. He concluded that though the results did not quite meet today's standard, .05 level of significance, analysis showed "significant dermatoglyphic differences between schizophrenics and normals when the former were individuals with positive family history for schizophrenia." It was, in short, a belated second endorsement.

CHAPTER 20

THE VISIONARY EXPERIENCE

W hile researching this book, I found few individuals still alive, other than my father, who had been associated with Huxley's experimental circle of friends. One, of course, was John Smythies, who had directed me to Fee Osmond, told me about Outsight and Eileen Garrett and Le Piol.

I learned one day when sitting in John's office at UC San Diego that my father and Humphry Osmond had something unlikely in common. I had just told John the story about my father's childhood fall on the ice and how neurological damage apparently changed him from the school's star reader into a dunce, until he adapted to his injury. At this, Smythies said, "Why, a similar thing happened to Humphry!"

Osmond's fall on the ice took place toward the end of his career, when he was director of the Alabama State Psychiatric Hospital in Tuscaloosa. After Osmond's fall, his wife had urged him to have it checked out, but Humphry said no because he thought he was fine. However, he was not fine, and the untreated cranial hematoma caused neurological damage affecting the frontal cortex in the area where reading comprehension occurs.

In contrast, my father's injury had taken place when he was young enough to develop compensation for loss of reading ability. Smythies said that at Osmond's more advanced age, though he was still able to carry on successfully through collaboration with colleagues, the damage did limit the work of his later career. Humphry

Osmond, born the same year as my father, in 1917, would die of cardiac arrhythmia at age eighty-six.

Figure 8. Humphry Osmond, 1965. Photo by Jane Osmond. Courtesy of the family of Humphry Osmond.

Figure 9. Aldous Huxley interviewed by an unknown journalist at the parapsychology conference in St. Paul-de-Vence, France, 1961. Courtesy of the Parapsychology Foundation.

When it came to my father's prospective lifespan, I already knew by 2001 that time was not on my side. Ever since I had returned from New York and Dad and I had started spending time together, he'd told me that when he thought he had lived his life fully he planned to make a choice rather than wait around.

I was hoping the rediscovery of the hand project might reawaken his old inventive verve, inspire him to stick around for a few more years, selfishly I suppose, so we could enjoy each other's company and make up for lost years. Then, one day in late summer of 2001, and unaware that within two weeks the nation would witness an outrage reminiscent of the Pearl Harbor attack, I said something like, "So, Dad, this is the era of the genome when science is uncovering biological indicators for so many things that were once a mystery. I think this would be a good time to revive the hand study and try to bring some of your insights to light."

"No," he replied. "For me that window has closed."

Nonetheless, I continued my research and interviews and followed leads and I kept a close eye on Dad, who insisted on living independently in an old stucco duplex that looked so much like a toy Spanish-style house that I called it his casita. He lived with few material goods, even though he could afford more. A sleeping bag. A desk. Two chairs. Some clothes. He had thumb-tacked a few photos on the wall—one of my stepmother, who had died in 1992, and a current one of me; an old photo of his childhood farmhouse and one of his modified Step-van camper, number three or four, parked by a creek under a canopy of billowing clouds in one of his favorite places, an oasis near Death Valley.

Then one day when he was in the hospital for some tests and I was straightening up his small apartment, I found the gun. The muzzle taunted me from its hiding place, but when I reached for the .38 I thought about what my independent father would do if I took it away. I concluded that he would just replace it.

Dad and I had become closer than at any time since the days when I assisted him with the Hand, back when we'd traveled around to take photographs of so many different types of people, when we drew a crowd to our booth at the hobby show, worked together in our closet darkroom under red lights over pans of developing images. We were closer now than then.

There had been that time, around two decades ago, when I had missed an opportunity to get to know him better, when I evaded his offer to teach me the art of healing. After that he did not ask again, and I figured that he had quietly retired from the whole healing endeavor. Then a few months later, around the time I found the gun, Dad took on his last subject, someone I'll call Sam. Dad had mentioned him in passing but said little about their relationship. One afternoon when I stopped by the casita to see Dad, I ran into Sam.

"So you're the daughter I've heard so much about," Sam said, pumping my hand, telling me how low he had been until Howard lifted him up with his technique and set him back on his feet. As I understand it, Sam had been a healthcare professional with a wife and children and a home in the suburbs until a mental breakdown overwhelmed him. Apparently, after several attempts at therapy failed, Sam lost his job. His family life deteriorated and his wife asked him to leave. That was when he moved into the city and rented a small place that happened to be near my dad's.

My father and Sam met by chance at a neighborhood coffee shop, probably the kind with a counter and a dozen stools, where guys order a solitary breakfast or pie and coffee in the afternoon, and where it's easy, if you're in the mood, to start a conversation. In this instance, it led to a stranger's admission of being confounded by his illness. My father obliquely introduced the subject of unconventional healing, and seeing interest he described his technique and the varied medical problems of people he had helped in the past. Sam said he was interested in being my father's subject. This meant undergoing the same kind of sessions I had witnessed years ago in the beige-carpeted rented office space in Calabasas, sessions where my father stood in silence swaying slightly, hypnotically, his hands suspended as if tracing the shape of the patient's second skin.

By this time months had transpired and Sam had improved enough to return to his career. He had stopped by Dad's casita to share the news of a promotion, and that's when Sam told me my father had saved his life and how he owed him the greatest of favors.

A short while later, on December 6, 2002, I wrote this in my journal: "Dad sprained his foot and needs mobility aids for two weeks but says he is ready to accept my help. He had a walker and a cane, and when I brought some of his favorite foods—a package of berries, a yam, a ripe avocado—he smiled encouragingly, I thought, so I told him: Soon you'll be good as new."

It was barely dawn the next morning when the doorbell rang at the house where I lived with my husband, Alan, fifteen miles away from my father's apartment. I threw on a robe, peered through the window, and saw a uniformed officer standing on my porch with a piece of paper in his hand. Robe rumpled, hair disheveled, I opened the door and heard him say, "Ma'am, you'll need to call this number."

"What number?" I asked, and he told me it was the LA County coroner's office. Shaking, I phoned and heard as if from underwater a voice asking me to come to Long Beach, to Dad's apartment, to make a positive ID.

When I got there I crumpled into my father's chair. They spared me the worst with a Polaroid. The assistant coroner offered condolences, then said, "You would be surprised how many older men living alone make this choice." Then she added, "Do you want to keep the gun?"

"No," I said. "Take it away."

It turned out that my father had not been alone in the casita before dawn when he acted on his decision. Sam had been there— the assistant coroner asked if I knew him, and I said yes—and it was he who had called the police, and he was waiting when they arrived, but he was gone by the time I got there. They did not suspect foul play, nor did I, because my father had warned me that when he was ready he would choose his day. He left a note for the police to call his daughter. Sam was not detained.

My father's body had been removed before I arrived. They had left now, the assistant coroner and her people, and in the aftermath I sat bewildered in Dad's big chair, stunned and desolate and waiting for the landlord, whom I'd been told was on his way. For though it seems like adding cruelty on top of grief, there are legal hazmat steps the relatives or the landlord must follow in connection with a suicide by handgun.

Before the landlord arrived, I tried to piece together what must have happened. Sam had apparently spent the night at Dad's request, and I was glad my father hadn't been alone. Behind a closed kitchen door, he pressed down on the trigger. Sam must have known the plan, his presence a few feet away a rare parting gift, the greatest of favors.

My father had reached for the gun in the small hours of December 7, 2003, the anniversary of Pearl Harbor. A declaration of war sixty-odd years ago, with its myriad ramifications, had led to the two epiphanies of his life: the mysteries of the hand and the mysteries of healing. Probably, he thought as he departed: *My daughter will understand. I won't need to explain in a note*. In a way, like his friend Aldous taking LSD on his deathbed, my father had scripted his own fate.

After his death, I temporarily stored Dad's ashes in a spare closet along with his camera equipment and mine. The following winter I stood on a desolate dirt road in Panamint Valley, just outside the Death Valley National Park, holding the interior plastic bag. I removed the official metal seal. When I opened the top and offered the first portion of ashes, the wind carried it back toward my face, so I turned around and poured while walking the opposite way. As I felt the bag lighten, I glanced over my shoulder and saw Dad's ashes lofting up the canyon like a pale oblong wing.

Approximately one year after my father's passing, Humphry Osmond died, on February 6, 2004. That month marked the fiftieth anniversary of the publication of *The Doors of Perception*. His daughter told me they scattered his ashes under the apple trees at Onet Cottage, the Osmonds' family home near Godalming in Surrey, a county in the southwest corner of England.

I returned to Laura Huxley's place in 2007 for the second time that year. A couple of weeks earlier I'd made a single short trip up

Beachwood Canyon to pick up a photograph after Laura had given me permission to make a copy of the wall-mounted 1963 photograph of Aldous Huxley's hand.

The first time I saw it, I had pointed out to her how this image had been so overexposed by the flash that half of the palm resembled a bald spot, unlike the well- lighted one my father had made in the 1950s. She gave me her ok to do a scan, and I said I would have my photo lab in Long Beach digitally bring forth latent content. Laura, though well into her nineties, was still interested in such curious matters as hidden patterns in the hand, which had been Maria's interest rather than hers.

On that day of mild weather, my second trip up the canyon in a month, I was back at the house on Mulholland Highway to return the original and give her a duplicate of the enhanced version she'd requested. Greeting me in the foyer, she asked if I would please hold her arm for assistance while we walked to a shaded area in the rear of the house. When we reached the terra cotta patio I helped her to her seat then, taking a chair facing the Hollywood sign on the hill, I sat across from her at a small table. In addition to a copy of the enhanced hand photograph, I had brought her a copy of the full set of Huxley-Osmond letters which, to the best of my knowledge, only existed in two places, my digital copies and the original black binder now safely back in the hands of Fee Osmond.

I explained that I had carefully read in sequence the hundred or so letters exchanged between Huxley and Osmond, and she asked me if I would read a few of them to her aloud. I had tagged some written by Aldous that seemed the most telling or poignant, and some that reflected a shared British sense of humor between the two men. I often smiled as I read them, and she kept encouraging me to read on.

And as I did, along with my own intermittent commentary, I told Laura how these letters had helped me understand that little-known circle of Huxley's friends. I mentioned how ironic it seemed that Freud's portrait was featured on the cover of *Time* in 1956, when at the same time the Outsight project, Osmond and Huxley's proposed psychedelic study, was losing its struggle against the Freudian and Behaviorist gatekeepers.

Laura picked up on this idea and said, "Aldous wrote about such things, about fashions in psychology. I think it is called 'The Oddest Science,'" Later on, after I had found a copy of the essay originally published in *Esquire*, I saw how Huxley talked about where psychology had taken a wrong turn by following Sigmund Freud and James Watson, the wrong turn Osmond had suggested to Aldous. After I left the house, Laura's personal assistant sent me an email to thank me for bringing the letters. She told me how much Laura had enjoyed hearing them.

But I need to step back a bit to the day when I first picked up the 1963 hand photograph. I was told when I arrived that Mrs. Huxley was not feeling well, so her assistant handed me a large envelope containing the photograph that had been removed from the frame on the wall. I wanted to examine it more closely once I reached my home, before taking it to my photo lab for a high-quality digital scan. And that is when I noticed. Not only had Aldous signed his name but someone had written a date on the back: November 16, 1963. Even now I am mildly astounded to think that when Aldous no doubt knew he was dying he would sign his name, in what became his final week, on a photograph of his hand. I take it as a message that, like his final session of LSD, the mysteries of the hand had deeply mattered to him.

Wafting skyward in the desert was my father's idea of a memorial service, and he wouldn't have wanted it any other way, but for a long while afterward I had a lingering need for closure. Then I found a way to honor him indirectly, four years after his death. I had seen an announcement for an event taking place on March 30, 2007, and had booked an early flight to San Francisco to attend a celebration honoring Huston Smith, a living treasure whose book *The Religions of Man* had made a great difference in my life and that of many others. I first read it after the summer of my moth encounter, when I changed my college major from journalism to philosophy.

Distinguished friends of Smith, practitioners of many faiths, gathered in Grace Cathedral that day—Tibetan Buddhist, Vedanta, Sufi, Jewish, Christian, Native American. They came from around the United States and from countries around the globe to offer words of inspiration and share memories with their honored guest.

But it was expression from the realm of no words that caused my spirits to take flight: brass gongs and the hum of Tibetan Buddhist chants, Native American drums and ascending wisps of smoke, a circle of full-skirted Sufi devotees clad in white, whose dance became a whirling vortex, and I was reminded of how physiological change can lead to altered perception and a glimpse through an otherwise closed door.

What took place inside Grace Cathedral seemed like a tableau from Huston Smith's own landmark book. We were a congregation of strangers gathered to honor this wise old man whose biography, the final chapter yet to be written, now appeared as if highlighted in a series of overhead slides. The audio track had been prerecorded by friends and, for some of the images, by the honoree himself.

At one point a black-and-white photograph of a two-story frame house at the end of a dirt road appeared on the large screen. It was accompanied by Smith's recollection of the day he spent with the Huxleys at their Mojave getaway, when he and Aldous had walked through the Old Testament–like setting, with its bleak expanse, and talked about the teachings of the Desert Fathers.

It would be half a dozen years before I had a chance to drive up that same dirt road, but when I reached the familiar 1880s facade framed by a stand of mesquite and opened the car door my senses became attuned to the pungent whiff of sage. The ranch house of white clapboard, then as now, sat surrounded by open land with its two stories rising up in the middle of nowhere.

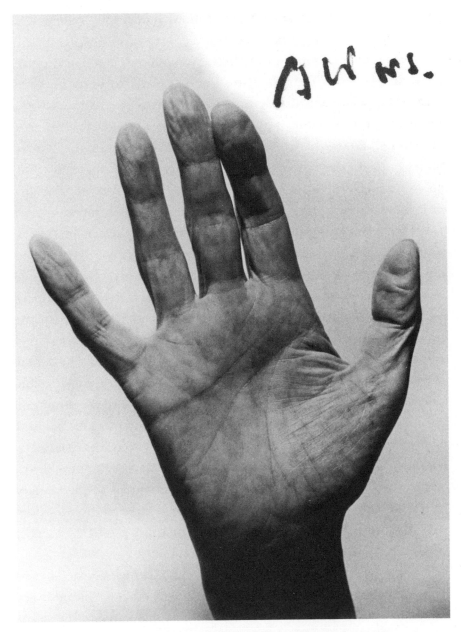

Figure 10. Huxley's hand photograph with his name signed by him one week before his death. Photographer unknown. Courtesy of Aldous and Laura Huxley Papers, Library Special Collections, Charles E. Young Research Library, UCLA.

This was no tourist site with guidebook and posted hours but a private home, and I'd tracked down the owners and explained my purpose. Huston Smith's words from his memoir came back to me. He'd said he couldn't believe his good luck to be standing here, and I felt a rush of gratitude on a warm Mojave day in the Aldous-significant month of May.

He and Maria had moved here around the time the United States entered the war, in late 1941. It now sits on five acres, formerly on forty, and I can see its appeal to Aldous, especially because of its once-full acreage. The vague enormity seems a metaphor for how his eyes saw the world. At the edge of the original property line stand romantic stone ruins of the failed utopian colony of Llano del Rio. Back when Aldous lived here, the colony and the ranch were separated by a natural irrigation channel. Proximity to this symbolic ruin alone might have prompted them to pony up the ranch's asking price.

Anne, the current owner, is well versed in Huxley lore. She led me through the gate of the high chain-link fence and invited me inside, where I was surprised to see the small living room until she reminded me that the house was built before the turn of the previous century. There was once an old wooden staircase, but it had been replaced by a wrought-iron spiral installed by Aldous and Maria's nephew Siggy.

From this cramped space, I could see why Aldous needed to add a structure for sleeping and writing. Anne took me to the two-room studio, built by a local craftsman but said to have been designed by Aldous. Anne still has a copy of the original plan.

Out back, I imagined Aldous caring for his vegetables and grapes. I knew grape stomping had gone on here, and bottles were stored in a wine cellar with a Wizard of Oz–style flap-covered entry. There had been cows for milk and cheese, tended by the Huxleys' groundskeeper, and there were still remnants of a livestock shed that reminded me of the old man said to heal both man and beast, who had served as a model for Huxley and Isherwood's unpublished screenplay about a healer.

The property line was once adjacent to a primary dirt road, but now the fat blue Los Angeles aqueduct slinks past and the main road is on the other side of the property: Pearblossom Highway, Route 138, a two-lane alternate route between LA and Las Vegas, and reputedly the most deadly road in California. Even half a mile away, Anne and her husband sometimes hear the screaming metal of a deadly crash.

November 22, 2013: A Personal Remembrance.

Yesterday marked the fiftieth anniversary of JFK's death, and thousands gathered at Dealey Plaza in the city of Dallas. In Los Angeles, where Huxley had lived for three decades and died on this same date, an exhibition called *Aldous Huxley and the Visionary Experience* had recently opened at UCLA for a three-month run. As soon as I heard about it I was on site the next day.

Reaching the UCLA Library Department of Special Collections, I walked down one flight and at the foot of the stairs found myself facing the exhibit poster showing Aldous in the 1920s, a young writer newly arrived on the international literary scene. Once inside, I found myself among photographs, books, and objects in glass cases; on the walls hung farmed portraits, and a monitor showed images recalling incidents that I had been writing about, and in many ways living with, for the past dozen years.

Here was an array of devices Aldous used to aid his damaged eyesight: pinhole glasses, his spyglass for distance, and his magnifier for close up.

A slide flicked by on a monitor, an ink sketch of a hand drawn by Huxley.

Inside a long case with other memorabilia lay a 1953 photo of him on a hill, captioned "photographer unknown," but I knew it was

Osmond, photographer and guide, who had snapped this image of Huxley with arm outstretched on a bright day in May.

A framed line drawing of Aldous by portrait artist Don Bachardy, who for decades had been the partner of Christopher Isherwood until the latter's death, showed a somber side of Huxley, with his head resting on his hand. This was the drawing that had appeared in the posthumously published edition of *Jacob's Hands.*

Don had helped me make the connection with Laura Huxley in 2001, and I saw him again in 2014 at a bookstore reception for *Hollywood,* his new book of portraits. A few weeks later I phoned to ask him what it had been like to draw Aldous Huxley.

"Nervous-making," Don said. "He was such a distinguished man, and his whole manner was so submissive and polite and sensitively tuned that I was unbalanced by it at first. I did four drawings of him in that one and only sitting from 1962."[1]

Of the four, I had seen two, and I liked this one because of the prominence of his hand. "His hands are in other pictures of him, too," Don explained. "There was another one Chris liked better than that one," meaning better than the head-resting-on-hand pose. I had seen that other one, too; it shows Aldous with his hand on his knee, looking to the side. "Chris thought it went deeper. It's not as much a close up. It was a certain mood Chris saw in the drawing that he thought was characteristic of Aldous, and he hadn't seen it captured in photographs of him. That was Chris's favorite. I think the drawing was Huxley's favorite of the four." So maybe Huxley thought the looking-away pose captured an unusual side of him too.

I asked him where the originals were now, and he said he had one at home, one was in the Metropolitan Museum in New York, one was in the National Portrait Gallery in London, and the one I liked, of Aldous with his head against his hand, was part of the UCLA Library Department of Special Collections.

The name chosen for the exhibition couldn't have been more

apt: *Aldous Huxley and the Visionary Experience*. The phrase became a touchstone after his 1953 psychedelic journey, and for his last decade it served as the centerpiece for many of his lectures at home and abroad. Because this theme was part of his repertoire, it raises the question of whether this was only in retrospect, or whether he familiarized himself with psychedelics anew?

The best answer appears to be that he took mescalin etc. (mescalin, LSD, and psilocybin, not counting a few minor experiments such as ololiuqui) about once a year, as he told John Chandos in a 1961 interview. Laura Huxley would later confirm that throughout the decade of 1953–1963 it was a dozen times in all.[2]

The visionary experience summed up his quest for perception, both ordinary and extraordinary. Huxley, of course, was nearly blind, and vision had more layers of meaning for Huxley than for most people. "A little known fact is that psychedelics can sometimes improve visual acuity temporarily," said Dr. Charles S. Grob of the Harbor-UCLA Medical Center, who is one of today's leading investigators in the field of psychedelic science. "Suddenly Huxley was seeing colors and objects, in part because his capacity to visualize was so impaired."[3]

To Aldous, the visionary experience was first of all a fact—a fact known by shamans and saints and often the dying and even ordinary individuals. It could be accessed by three means: spontaneously or through psychophysical methods or by ingesting certain harmless (or nearly harmless) chemicals. To Huxley, the visionary experience and, beyond it, the mystical experience, were not cases of pathology or symbols or myths but supremely valuable empirical facts of human existence.

In one of his last talks, given in 1962 at Monterey College on California's central coast, Huxley referred to the highest expression of this state of consciousness, the full-blown mystical experience: "This is felt to be of immense importance," Huxley said, "and it does, in fact, help many people to change their lives in significant ways."[4]

Figure 11. Drawing from life by Don Bachardy, August 2, 1962. © Don Bachardy, 1962. Used with permission.

CHAPTER 21

THE RETURN OF PSYCHEDELIC SCIENCE

A certain phrase often mentioned in any gathering where the topic is psychedelic science comes from Huxley's final novel, *Island*. Ever the wordsmith, he borrowed one component from Indian philosophy (*moksha*, referring to emancipation or freedom from the cycle of birth and death) and added medicine, so that moksha-medicine becomes a balm for treating afflictions both physical and spiritual. A mushroom-based hallucinogen, this elixir is not consumed with casual disregard on Huxley's fictional island. I recently reread the novel and was surprised to see how Huxley gives us a prescient glimpse of moksha-medicine at work today.

Eager to snag a good seat, I arrived early in Pasadena to hear a talk about "Psychedelic Science: From '60s Counterculture to Modern Medicine," a forum sponsored by Southern California Public Radio station KPCC on a warm September night in 2014.[1]

I would soon sense, by catching the number of knowing expressions, that most audience members had experienced these substances firsthand. They, or rather we, knew about distortions in time and perception, how these drugs gave rise to visions, how they often induced profound feelings of interconnectedness.

Clearly, generational change is bringing a more accepting attitude toward substances banned for forty years, and the proof is that officially sanctioned human research is back. I had been following news reports and Google alerts, but unlike the '60s, with its troika of mescalin, LSD, and psilocybin, the information streaming in needed translation, a guide for the psychedelically perplexed. I had questions. What about the dizzying roster of substances currently called psychedelic drugs—how did they fit the picture? How did a therapeutic session relate to neurological research, tell the imaging story of powerful substances acting on the brain?

Having done explanatory journalism in a former life, I aimed for immersion. This seemed about to begin when I stepped inside the theater-like ambiance of the "Psychedelic Science" forum, where blue lights evoked the dreamlike feeling that I was about to dive into an indigo pool. The setting made sense, with its moody tint and a soundtrack from before mind-altering substances were banned, like Jackie Wilson's 1967 pop hit "(Your Love Takes Me) Higher and Higher."

Wilson's song hit the charts three years before psychedelic research went into full hibernation. That happened after LSD, psilocybin, and other drugs were banned under the sweeping prohibition of the Comprehensive Drug Abuse Prevention and Control Act of 1970. An amendment to the law established five categories or schedules, capped by Schedule I for drugs deemed to have no medical purpose, unsafe even under supervision, and with a high potential for abuse.

This designation partly explains why the operative phrase on this night was "modern medicine." This panel was not mainly about psychedelics for expanding creativity (though the Beckley Foundation in England is funding a pilot study on creativity with University of California, Berkeley), nor about inducing mystical experience (though a three-part psilocybin study was conducted from 2001–2006 under the auspices of the Johns Hopkins School of Medicine). Nor did talk turn to recreational use, because the first matter of business in order to utilize a Schedule I drug is to demonstrate medicinal value.

Or, prove it once again. Before such prohibitions went into effect, circa 1970, mental health professionals had published an estimated one thousand clinical papers involving tens of thousands of participants. Such research often fell below today's methodological standards, being mainly anecdotal accounts rather than double blind, randomized, and placebo-controlled studies with results tracked using specific questionnaires or psychological measures called inventories.[2]

Despite design flaws, early studies showed good outcomes for patients suffering from afflictions ranging from depression and anxiety to alcohol addiction. Treatment with psychedelics was beneficial for improving the mental state of subjects suffering from deep trauma, as well as patients facing terminal illness. The two most recent studies, described in current terminology, examine psychedelics as a treatment for anxiety in advanced-stage cancer patients and look at psychedelic-assisted psychotherapy for posttraumatic stress disorder (PTSD), and these two studies now lead the way in the rebirth of psychedelic science research in the United States.

LSD was formerly the preferred drug for such investigations. Holdover baggage from the '60s means LSD is currently less likely to win research approval from US government agencies, though a LSD-assisted psychotherapy study with advanced cancer patients in Switzerland was recently completed,[3] as was a pilot study in the United Kingdom.[4]

Enter the lower-profile alternative, psilocybin, the active compound in magic mushrooms and a substance used in several categories of psychedelic research today. Another relative newcomer increasingly in favor is MDMA (3,4-methylenedioxy-N-methylamphetamine), which may seem an unlikely choice considering its dance-club reputation as Ecstasy. When I attended an event not long ago I overheard one difference described this way: "With LSD once you take it, you're on your way. MDMA is gentler and easier to control."

Similar in effect to marijuana, MDMA is structurally akin to mescalin but without its perception-altering qualities. MDMA, under the name Adam, became available in the mid-1970s and was found to induce an empathetic response beneficial in the therapist-patient dynamic. Consequently, it was prescribed to patients and clients by mental health professionals and, prior to restriction as a Schedule I drug in 1985, showed promise in psychotherapy for conditions ranging from phobias to depression.[5]

The two compounds moving the field of psychedelic medicine forward are MDMA and psilocybin, according to Charles S. Grob, MD, chief of Child Psychiatry and professor of Psychiatry and Biobehavioral Sciences at UCLA's David Geffen School of Medicine. Grob was the first researcher in decades to lead an approved study with MDMA. That pilot study was conducted in the mid-1990s,[6] though much of his subsequent work has been with psilocybin. Grob is also co-founder of the New Mexico–based nonprofit Heffter Research Institute, which supports research of classical hallucinogens and related compounds.

If MDMA and psilocybin are the most promising drugs in this rapidly expanding field of research, then the type of study farthest along in the approval process turns out to be psilocybin-assisted therapy with advanced cancer patients. This study originated with Grob's team at UCLA,[7] and where it stands today gives an idea of how the process works.

"After our study began, two other research groups began studies at New York University and Johns Hopkins," Grob told me in a telephone interview a few weeks before the Pasadena event. "They have just finished the treatment phase, and now we have to wait for a six-month follow up." That would take it into 2015. "Then there will be analysis of their data, a pooling of data from our studies, and we will submit a proposal for a multisite grant."[8] Results involving a larger population of patients is likely to bring psilocybin closer to FDA approval as a treatment.

Such approval would not make psilocybin a prescription for the masses like Prozac nor sanction a last trip like Huxley's (nor a dream-state exit like the Edward G. Robinson character's parting in the 1973 film *Soylent Green*). The application would be specifically for assisted psychotherapy in a medical setting during the period between diagnosis and the last months of life.

A reference to end-of-life and psychedelics naturally brings up the Huxley connection, because he left his mark in both literature and biography. "Huxley described it before any studies were done," Grob told me, noting that the pioneering study by Eric Kast of the Chicago Medical School[9] came after Huxley's last book, adding to Huxley's reputation as a futurist. "Huxley wrote about this in *Island*," Grob said. "He walked the walk when it was his time."

Whether applied to improve the last months of life or interrupt the grip of addiction, this field of old-is-new medicine is not about downing a pill and feeling better. Psychotherapy is an essential component in the process. In psychedelic-assisted therapy the drug opens the mind, which increases the chance of bringing about behavioral change. The hoped-for outcome may be smoking cessation, breaking free of horrific recurring memories of rape or a firefight in combat, or escaping an obsession or deep depression. On the procedural level, a course of treatment might involve several three-hour preparatory sessions before drugs are administered, subsequent psychotherapy sessions, and longer-term follow-up.

Terminology referring to substances in this field can be confusing. The word psychedelic remains the popular, eyeball-drawing, often-preferred-by-the-media term, though usually the preferred medical term is hallucinogen. Another term often used in psycho-spiritual studies with hallucinogens is *entheogen*, meaning god- or spirit-facilitating. Pinning down terminology for mind-altering substances dates back at least to German pharmacologist Louis Lewin's five categories of psychoactive plants and drugs set down over a century ago, when his list consisted of euphoriant, inebriantia, exitantia, hypnotica, and phantastica. Lewin had hallucinogens pegged under the last category.

In more recent times, categories and usage have been in flux since the mid-1950s, and even today debates arise over how to categorize MDMA, which has been variously called an *empathogen*, for its effect of creating an empathetic state of mind, or an *entactogen*, for its ability to assist in recovering repressed memories.[10] One such debate surfaced in 1956, of course, when Huxley and Osmond exchanged letters about needing an umbrella term for a burgeoning number of psychoactive drugs. Today, many opponents of these drugs call them psychotogens because in the medically predisposed wrong hands they can trigger prolonged psychosis.

Whether you call them psychedelics or hallucinogens (or psychotogens), because these are Schedule I drugs the key to reopening a legal door is through the process of federally and locally approved clinical studies. Even gaining approval for a pilot study would have been out of the question until just over a decade ago. "As time goes on," noted Grob, "I am hearing less controversy, more resonance."

Generational change has contributed to a shift at the Food and Drug Administration and the Drug Enforcement Agency. As personnel who experienced the 1960s are promoted to higher positions, their approach to policy is likely to differ from those under the sway of the 1970s War on Drugs.

Rising healthcare costs provide another opportunity for change associated with two large subgroups: aging baby boomers facing end-of-life care and thousands of returning veterans diagnosed with long-term posttraumatic stress disorder, or PTSD.

PTSD is a centerpiece of clinical studies funded by the Santa Cruz, California-based Multidisciplinary Association of Psychedelic Studies (MAPS), whose founder and director, Rick Doblin, has a doctorate from Harvard in public policy. MAPS was established around the time MDMA became illegal in the mid-1980s, with the mission of supporting psychedelic research and promoting changes in drug policy. It is the only organization specifically funding research leading to clinical trials of MDMA-assisted psychotherapy, though it funds studies with other psychoactive drugs as well. The nonprofit organization's stated priority is to make MDMA a prescription drug by the year 2021.

Among the MAPS accomplishments so far is a study of MDMA-assisted therapy for posttraumatic stress disorder with US veterans in South Carolina, and similar studies are underway in Vancouver and Israel. One of the latest MDMA studies, approved in November 2014 and funded by MAPS, is a study of MDMA-assisted psychotherapy for anxiety associated with life-threatening illness, to be conducted with eighteen subjects in Northern California.

MAPS also had a presence at the "Psychedelic Science" event in September of 2014, when a video sponsored by MAPS showed interviews with several PTSD patients who had undergone MDMA-assisted psychotherapy. One was a young mother named Rachel, who had been a victim of childhood sexual assault. She recounted how she unsuccessfully sought help for her PTSD until MDMA-assisted psychotherapy enabled her to dredge up painful memories and recast them in a new light.[11]

A dramatic narrative such as Rachel's adds a human face and a persuasive power to the story of psychedelic research, especially in an era of sharing via Facebook, Twitter, and YouTube. Less alluring is the practical side of research, the data essential for a clinical study, involving inventories or questionnaires that can be statistically quantified. A personal account, which may have the effect of a testimonial, conveys the message that psychedelic therapy deserves a second look, even though one could say that an anecdote speaks for a single subject. By contrast, data based on larger numbers of participants is essential to advance a clinical trial.

I was drawn in, actually taken up short, by one such personal story. I first read about it in an article in *Discover* magazine in which journalist Linda Marsa quoted a woman named Annie who had been a participant in Charles Grob's 2004–2008 study with terminal cancer patients.[12] I learned that Annie's ovarian cancer, which had gone into remission, had returned; I read how, through the process of psilocybin-assisted therapy, she had somehow unclenched, let go of her fear, and opened up to experiencing greater joy with loved ones in the time she had left. Then I found a video of Annie's interview on the Heffter Research Institute's website, and I heard her say, "It connected me with the universe."

Annie's story hit home because I am an ovarian cancer survivor. At the time of this writing it has been a dozen years since my surgery, but when I read how Annie's cancer had "come roaring back" then I saw the sensitive interview, well, it gave me a queasy taste of my own mortality. I know that when my time comes near I would want this consolation, though it is not legally available yet. I guess that makes me an advocate with a dog in the fight.

This brings up another reason why the attitude toward psychedelics is changing. Your mind can be changed when you personally know, or have heard about, someone who has been helped through psychedelic therapy. This seems to be a persuasive factor in the states where voters have approved medical marijuana.

The current reevaluation of psychoactive substances comes after a forty-year period of dormancy, depending on how you look at the calendar. Some people say it has been four decades since a new study was approved. Or, viewed another way, there were twenty years between the Controlled Substances Act of 1970 and the next legal psychedelic study, which began in 1990. That was when psychiatrist Rick Strassman at the University of New Mexico School of Medicine designed an experiment with the hallucinogen DMT (N,N-dimethyltryptamine). DMT is an active compound found in ayahuasca, an Amazonian psychedelic tea attracting both scientific and popular attention today.

Strassman, who conducted his research with sixty volunteers, was not looking for a therapeutic benefit but was more interested in how DMT induces a psychospiritual, or mystical, experience. His study ended after five years without advancing further, apparently due to various factors at the sponsoring institution.[13]

Taking up a similar research question in 1993, Charles Grob conducted a private study in Brazil of ayahuasca in the UDV church (Centro Espírita Beneficente Unão do Vegetal), for which drinking the psychedelic tea is part of a sacramental ritual.[14]

Since then, and especially since 2000, a number of clinical studies have been conducted in the United States, many spanning several years and continuing today. Some reprise work interrupted decades ago. An example is the previously noted Harvard–Marsh Chapel Experiment in the early 1960s, examining psilocybin as a catalyst for psychospiritual or mystical effects, a study conducted by Walter Pahnke and Timothy Leary. A legacy from that project is a measurement used today called the Pahnke-Richards Mystical Experience Questionnaire. The MEQ, a self-report tool, assesses seven areas of mystical experience, such as transcendence of time and space, and a sense of the unity of all things. Volunteer responses are rated on a 6-point scale.

A subsequent 2001–2006 version with psilocybin was the one conducted at Johns Hopkins University School of Medicine, headed by Roland Griffiths. In both cases many of the participants reported profound spiritual experiences. In the first round of the modern study, which was more carefully managed than its predecessor, an analysis of responses from the thirty-six volunteers tracked fourteen months later showed that 58 percent rated it as among the five most personally meaningful experiences of their lives.[15]

Another once-promising study looked at psychedelic-assisted therapy for psychosocial distress in terminal cancer patients. The first study, under Eric Kast of the University of Chicago in the mid-1960s, was an outgrowth of an investigation of LSD as an analgesic for patients with advanced cancer, but Kast discovered that his subjects were experiencing a transformation in how they viewed the end of life.[16] The last LSD-assisted terminal-cancer study took place in the early 1970s at the Maryland Psychiatric Research Center, led by psychiatrist Stanislav Grof, whose results informed the design of the Charles Grob–led psilocybin end-of-life study conducted from 2004–2008.[17]

Psychedelic-assisted treatment for alcohol addiction was one of the first therapeutic applications in the 1950s, led by Humphry Osmond and Abram Hoffer in Canada. Currently, a study investigating the therapeutic potential of psilocybin in alcohol dependence is underway at the University of New Mexico; another addiction study, this one for smoking, is being carried out at Johns Hopkins in Baltimore, and positive results could improve support for studies of hallucinogen-assisted therapy for other addictions.

Among outpatient psychiatric disorders, one of the most common is obsessive-compulsive disorder. The Heffter Institute has sponsored an initial psilocybin-assisted study at the University of Arizona, looking at obsessive-compulsive disorder,[18] Though results seem to warrant further investigation, no more studies are underway as yet.[19]

A recently-launched study in Los Angeles is investigating MDMA-assisted therapy for social anxiety in patients with adult autism, the latter a study funded by MAPS and led by Grob and his team at Harbor-UCLA in conjunction with Stanford University.[20]

After citing examples showing that psychedelic research is on the rebound, I would like to return to the forum in Pasadena—a city associated with art museums, a splashy New Year's Day parade, Fuller Seminary, and the California Institute of Technology, or Cal Tech, as it is better known. In short, a suitable locale for a chat about drugs, with implications for creativity, spirituality, and therapy, as well as the weighty subject of the science of the brain.

When it came time for the Q&A, audience members asked about clinical trial prospects for other psychedelic substances. Hands went up, names rang out: ayahuasca, ibogaine, ketamine. Because of research restrictions, all three are mainly studied abroad at present. MAPS is the primary sponsor of studies in Mexico and New Zealand involving the controversial drug ibogaine, derived from a West African plant, as a treatment for opioid addiction. Funding from MAPS and the Heffter Research Institute are supporting research into the effects of ketamine, an anesthetic with hallucinogenic properties being studied in St. Petersburg, Russia, as a possible treatment for heroin addiction. In another application, intravenous ketamine therapy is now available as a treatment for depression in psychiatric practices in San Diego and Los Angeles.

Unlike these two compounds, the South American tea ayahuasca is almost a household name, thanks to its popularity among psychedelic tourists heading to Peruvian shanties and retreat centers. Some would not call it psychedelic but by the more narrow term *entheogen*, meaning that it induces a mystic-like effect. Ayahuasca has a long history of ritual use in initiations and healing ceremonies. Celebrities, whose every move is splashed across print and online media, number among those taking the ayahuasca trail to the Amazon.[21] The psychedelic concoction, ingested under the supervision of a shaman guide, has been described as a wrenching ordeal bestowing beatific, though sometimes hellish, visions. For the fortunate, it is said to open the mind to personal revelations.

I first heard about ayahuasca eight years ago, on the day I met Professor John Smythies, who had been a catalyst for Huxley's milestone mescalin trip and is one of the last survivors of the era of original psychedelics research. A mutual friend named Bill had arranged the introduction and offered to drive us to Smythies's office at the Center for Brain and Cognition at the University of California, San Diego, which was an hour and a half drive away from where I live.

When at first Bill pulled up in his car, I was surprised to see another passenger inside who explained that he also wanted to meet Dr. Smythies. While we drove south on the San Diego freeway, through a barren stretch bordered on both sides by Camp Pendleton, Marine helicopters chuffing overhead, Bill's friend said he had just returned from Peru. He went on to describe his recent ayahuasca experience. It was not his first time. He had developed a relationship with a shaman, and every other year he returned to undergo what he called the initiation.

Bill followed this fresh-from-the-Amazon story by recounting one of his own earlier psychedelic experiences. My turn was next. Despite my journeys with LSD, I felt more inclined to tell them about my spontaneous experience, of how, when my eyes met the sepia pattern on a homely moth's wings, it had erased the present and carried me out of time. Bill's friend turned toward me, sitting in the backseat, and said, "That is all you need." But afterward my thoughts drifted back to my half-dozen LSD trips and memories of surfaces pulsing and sounds sparking colors and, above all, the oddly intimate and both frightening and fearless encounters with the wildlife of Golden Gate Park.

Today, the psychoactive tea seems to provide a bridge for exploring old and new questions. Grob's study a decade ago on behalf of the Brazilian religious order UDV showed that young people who participated in the ayahuasca rituals of the church were as well-adjusted as a control group of peers. Grob's current study, recently approved by the UCLA institutional review board and taking place in Peru, is expected to reveal more about the backgrounds of who goes to the popular ayahuasca retreat centers and why they go, whether the experience met their expectations and intentions, and whether there were good or bad effects.[22]

Without funding, however, such studies of the healing and other potential benefits of psychedelics would not take place. Clinical and preclinical research is expensive, with little in the way of federal or state grants available for work with Schedule I drugs (exceptions include some addiction research). Medical marijuana appears to be paving the way, however. MAPS received an unprecedented $2 million grant in December 2014 from the state of Colorado for a pilot study of marijuana-assisted therapy for symptoms of PTSD in US veterans. This looks like an opening for more than medical marijuana, or as Doblin of MAPS recently said, announcing the study, "The Veterans Administration is willing to collaborate with MAPS on PTSD-MDMA research."[23]

Charles Grob also believes that psychedelic medicine will eventually go mainstream, or as he said at the Pasadena forum on psychedelic science, "I think in ten to fifteen years it will be an accepted part of psychiatry and medicine, and there will not be a problem with funding. I am optimistic."

Major foundations largely stay clear of supporting projects involving these substances, as Huxley and Osmond found out in the 1950s when they sought grants from the Ford and Rockefeller Foundations in support of their own psychedelic project. But the current funding story has an upbeat though poignant sequel. It concerns the Rockefeller Brothers Fund and Richard Rockefeller, a great-grandson of John D. Rockefeller Sr. Long a generous donor to MAPS and a physician himself, Richard Rockefeller was interested in sponsoring psychedelic-assisted therapy for trauma survivors. Though he unfortunately perished in a private plane crash in June of 2014, Rockefeller made provisions for support of the work of MAPS to continue after his death.

Along with MAPS, of course, another major organization, in terms of seeking and granting funding for psychedelic research, is the Santa Fe, New Mexico–based Heffter Research Institute, cofounded by Charles Grob and David E. Nichols of Purdue University. The Institute was named for German pharmacologist Dr. Arthur Heffter, who in 1889 showed that mescalin was responsible for the psychoactive properties of peyote. Both MAPS and Heffter work across borders, as does the Beckley Foundation, based in Oxford in the United Kingdom. Under its founder and director, Amanda Feilding, the Beckley Foundation supports collaborative research with European as well as American university medical centers, including UC Berkeley, UC San Francisco, and Johns Hopkins University.

One of the Beckley Foundation's special interests is gaining a greater understanding of the effects of psychedelics by means of brain-imaging technology, which involves tracking cerebral blood flow of subjects who have been given a psychoactive drug.[24] It turns out that two of the leading experts disagree on what they see. Their debate brings to mind a pet theory of Aldous Huxley's.

Classic hallucinogens work on the brain, at the level of single neurons, by stimulating serotonin 2A receptors, as Robin Carhart-Harris and his colleagues recently explained in a special issue of the British journal *The Psychologist*.[25] Carhart-Harris notes that most of us picture neuroscience at the system level, in which a proxy for brain activity is seen in blood flow to particular areas of the brain. What we might see pictured in the media in a graphic (such as illustrated in red and blue) are three-dimensional volume pixels, called voxels, that have been visually translated into an activation map.

On the system level, how hallucinogens work on the brain is not understood, but one theory from the 1950s still has a kind of historical credence. In Huxley's view, as expressed in *The Doors of Perception*, the everyday brain acts like a filtering mechanism to reduce or organize input in a way useful for everyday decisions. In this view, substances like mescalin, LSD, or psilocybin override the filter, giving rise to what Huxley called Mind-at-Large. If this is the case, then in terms of magnetic imaging a mind-manifesting substance should produce an abundance of brain activity.

This happens to fit one neuroscientist's theory, though at first it sounds like the opposite. Robin Carhart-Harris and his colleagues saw in their subjects decreased brain activity in structures important for "integration and routing in the brain" meaning managerial function. If those practical hubs (thalamus, posterior cingulate cortex, and medial prefrontal cortex) register a decrease in activity then this more or less Huxlean filter-override could account for "an unconstrained mode of brain function."

On the other hand, Franz Vollenweider, a neuropsychopharmacologist at the University of Zurich who also sits on the Heffter board, sees it another way. Vollenweider saw not a decrease but an increase in activity in those same management-oriented areas of the brain, though the difference between the two experiments could have been caused by a variance in dosage or how dosage was delivered, orally in one case and by injection in the other.[26] For the time being, lacking agreement over how classic hallucinogens work on the brain, we do not know if Huxley (or philosopher Henri Bergson, who originated the idea of brain as a filter) was right all along.

The practical question is what happens if psychedelic-assisted psychotherapy, whether near the end of life or for treating PTSD, reaches the FDA approval stage and becomes a prescription drug? Or as one audience member in Pasadena during the Q&A session phrased it, "What would a treatment model look like?"

"It would not be just any health provider," Grob replied. "There would be rigorous training, supervision, and oversight. Otherwise there would be a risk of it spinning out of control."

CHAPTER 22

BESIDE PSYCHEDELICS

Imaging technology reveals parts of the brain activated during certain mental processes, resulting in hypotheses unimagined by those who attended Aldous Huxley's Tuesday night experimental salons in the 1950s.

Studies of neural activity during meditation may help explain psychospiritual experience but also pose new and fruitful questions. A study around a decade ago scanned the brains of Buddhist monks who had undergone thousands of hours of meditation practice. Imaging results showed activity in areas associated with positive emotions such as compassion.[1]

A study by neuroscientist Andrew B. Newberg in the area of neural plasticity, how our behavior modifies our brains, captured seemingly permanent changes in brain activity after subjects had practiced meditation for only twelve minutes a day for eight weeks.[2] Imaging might reveal the neural activity involved in the sense of non-duality, or unifying consciousness, involved in a mystic-type experience. A recent area of investigation by Dr. Josipovic Zoran, a research scientist at New York University, looked at the so-called default brain activity in Buddhist monks. The default brain appears to be involved in recognition of the self, how we know who we are, and how that sense of self might be overlaid in deep meditation by a feeling of interconnectedness with the universe.[3] Meditation as the path of psychospiritual practice is the likeliest state for such a study, though the experience of interconnectedness can occur spontaneously (not practical as a target for fMRI tracking), and may be catalyzed by ingesting a hallucinogen (which *is* a subject of fMRI study).

A more elusive question—what happens to consciousness in the first few minutes after the brain ceases functioning—was of literary and personal interest to Huxley. He agreed with his correspondent and friend professor John Smythies, who maintains that the mind is not identical with the brain. Huxley also entertained the possibility of a posthumous state, one he portrayed as a transitional phase in his novel *Time Must Have a Stop*, spoke of in his visionary experience lectures, and expressed in practice when during her last hours he read Maria passages from the *Tibetan Book of the Dead*.

A patient is considered clinically dead when the brain's blood supply and function ceases, which is said to be twenty to thirty seconds after the heart stops beating. Nonetheless, there are numerous anecdotal accounts of patients recovering consciousness, the so-called near death experience (NDE), which often involves having visions, witnessing a white light, and sometimes recalling spatially puzzling out-of-body experiences.

Because resuscitation is involved, these experiences invariably take place in a hospital setting. A recent study, published in the professional journal *Resuscitation*, reported results of a multisite study spanning six years and involving over 2,000 patients who suffered cardiac arrest. Of these, 330 survived. Though clinically dead, almost half retrieved memories from the time span when the heart and brain, and therefore the senses, were not functioning. In one verified report, a patient officially dead for three minutes clearly recounted details about the activities and events that transpired in the hospital room during what should have been, for the patient, a perceptual void.[4]

The notion of many paths leading to higher levels of consciousness, not only through psychedelics, but also by means of meditation, yogic practice, and hypnosis, appears throughout Huxley's writing.

These days, an echo of the experiments carried out by Huxley's Tuesday night group continues in work being done by several groups, some old and some of more recent vintage.

The Parapsychology Foundation carries on the work of Huxley's friend the medium Eileen Garrett, whose respected international conferences at Le Piol in the south of France in Huxley's day touched on topics ranging from nontraditional healing and psychic phenomena to psychedelics.

The Rhine Foundation carries on the work of Duke University parapsychology lab founder and Huxley friend J. B. Rhine, who in the 1930s popularized the term extrasensory perception (psi). Both foundations fund research, award study grants, and provide online repositories for sharing information about anomalous experiences.

Several new organizations dedicated to exploring states of consciousness were established after the 1970 ban on hallucinogens. One is the Institute of Noetic Sciences (IONS), a foundation established in 1973 and focused on a scientific approach to studying anomalous experiences ranging from nontraditional healing to remote communication.

I recently paid a visit to the IONS retreat center, located near its headquarters in Petaluma, California, and while there I noticed on a bulletin board that the University of California, San Francisco, Medical Center had recently held a faculty development conference at this venue. Until a few years ago, chances are the representatives of the medical establishment would not have imagined a faculty gathering taking place here.

No doubt an effect of the psychedelic ban was a renewal of interest in psychophysical techniques practiced long before hallucinogenic plants were synthesized in the lab. In the autumn of 2014 I heard an unexpected reminder of this slow but time-honored approach to altered states from MAPS director Rick Doblin. Though MAPS is dedicated to research and public policy involving psychedelics, Doblin acknowledges alternative paths, or, as he said at a MAPS-sponsored seminar in November of 2014, "One mistake of the '60s was to say psychedelics is the only way. I think that is false."[5]

One of the alternatives Doblin alludes to is a nondrug technique developed by Dr. Stanislav Grof. Grof and his late wife Christina developed a method called Holotropic Breathwork, which incorporates ancient spiritual practices of yogic breathing, meditation, and what might be called sensory deprivation.[6]

Grof, a medical doctor who also has a PhD, began his studies and research in Prague, Czechoslovakia, then became an assistant professor of psychiatry at John Hopkins University School of Medicine in Baltimore, which led to his role as lead researcher in one of the last government-sanctioned studies of LSD in the United States. When that era ended, Grof developed Holotropic Breathwork as a nondrug technique for inducing a visionary experience.

I had experienced two of the three methods—the spontaneous and the psychedelic, but not the psychophysical—so I decided to participate in a Breathwork session. It took place one crisp Saturday in November 2014, when my day began by driving north on the 405 freeway, branching over to Pacific Coast Highway, then making the turn into rustic Topanga Canyon, where I followed the twisty two-lane road through a slash in the Santa Monica mountains.

Reaching the community center, I saw that its interior, with its dark overhead beams, had been adapted for the task at hand. A purple wall hanging with a two-foot-high Sanskrit syllable hung over a large window, muting the sun. Four candles flickered in the center of the room beside a vase of long-stemmed blooms. A river-rock fireplace sat idle yet exhaled hints of old wood smoke.

About fifty of us had gathered to spend a day turning inward on a drug-free vision quest that for some, we were told, might evoke a kind of therapeutic release, a healing. We sat in a large circle to hear instructions then paired up, because the idea is for one to keep an eye on their partner, the roles reversing in the afternoon. Vigil is needed because bad trips can happen in such intense practices, even without drugs. Consider, as Huxley sometimes cited in his talks and essays, the drug-free visionary torments of St. Anthony isolated in his cave.[7]

Before the journey began, those of us on the mats donned eyeshades to disable normal vision and override a human's most taken-for-granted sense organ. It occurred to me that Huxley would appreciate how this technique was another way of bypassing what he called the practical "filter" of our everyday mind, an alternate way of opening the door.

Lying flat on my mat, I was about to plunge into sensory deprivation, crossed with sensory overload powered by extreme breathing. Shortly after the music began, my ordinary perception succumbed to an overpowering medley of rhythmic, sometimes eerie, sometimes not quite human, often pulsating, sounds.

The initial intense drumming seemed as if spirits of ancient tribes washed over me. Behind my eyeshade I saw pale grey wings perforated above the tips of the feathers stretching across the screen of my mind. When the background sounds changed, other shapes emerged: a deep blue aperture swelled invitingly but I was prevented from entering; to my right, a welcoming path covered with ochre leaves appeared, but I changed direction toward a stream that aroused in me a desire to merge. I breathed fast and deep, and when the everyday mind tried to retake command I pushed it aside.

Hours by the clock seemed like minutes. Another aperture with chevrons of color streaming to the sides turned into glossy black feathers supporting a raptor's head with a piercing green eye. That should have instilled fear but it filled me with joy.

After what seemed a spell of calm, another shape appeared. It was the public pool where my father, almost disastrously, taught me to swim. He had insisted that I learn by his sink-or-swim method of stepping off the mid-height diving board. That childhood terror came back, and so did the day of my father's suicide when I had been too stunned and numb to grieve. That day on my mat, I cried from grief and forgiveness.

Sounds of streams, night creatures, and birdsong guided us back for a soft landing, and no one left the candle-lit space until after sufficient time for decompression. As I drove home that night, I thought of how this compared to a session with LSD, of the profound sense of morphing and merging. To me, this day had been a replica, of sorts, on a smaller scale, and I was not disappointed. I also wondered about the mysterious, mind-opening healing process. I had recently read an interview with Stan Grof about his theory of bioenergetic blockages as possible sources of pathologies, an idea central to Chinese medicine and possibly at play here. It occurred to me that if psychedelic-assisted therapy opened the mind and was somehow able to help patients with buried traumas or depression or cluster headaches or many other afflictions, why shouldn't a mind-altering technique like this work in principle, too, or for that matter my father's healing technique, through the silent and hovering movement of his hands.

The inhabitants of Huxley's island effect healing with hypnosis, ritually take the moksha-medicine, and practice meditation. They master time-honored yogic positions, including tantric yoga with its focus on the sexual concentration of the life force, and they honor the pattern language of the body and what it reveals.

Clearly Huxley did not reject the body in favor of the mind, and maybe one could say he was comfortable with parallels without needing connecting lines. He thought the body gave clues to the mind and the mind to the body. His quest for perception was not only about visionary experience but also had a physical dimension he explored through practice, over the years, of such methods as the Alexander Technique and Iyengar yoga, and by espousing William Sheldon's theory of the somatotypes (endomorph, ectomorph, and mesomorph) as a source of interpreting human nature. On Huxley's island of Pala, children are taught what to expect of people whose temperament and physique are unlike their own.

A small step from this theory of patterns in the human form takes us to the idea that who we are and our place in the universe may be reflected in and through our hands. In 1925, early in Huxley's quest for perception, he proposed this idea in his novel *Those Barren Leaves*, through the voice of the character Mr. Calamy:

> And I believe that if one could stand the strain of thinking really hard about one thing—this hand, for example—really hard for several days, or weeks, or months, one might be able to burrow one's way right into the mystery and really get at something— some kind of truth, some explanation.[8]

Burrowing for a deep reading of the body largely takes place today in the fields of medical and security technology, both motivated by a practical quest for secure identity. Scanning technology is employed to uncover patterns naturally found in the body. This field, called biometrics, is described as the measurement and analysis of unique physical or behavioral characteristics, even including perspiration levels detected in an airport security line.

Sensors and scanners with high-resolution, multispectral technology glean layered details of fingerprints, hand-vein structure, and retinal patterns all as biometric identifiers, sometimes using two or more together in a more secure multiple mode.[9] Facial features provide facial recognition, and among the more unusual approaches to identification is a recently proposed biometric print of the tongue. Such research is mainly funded by government agencies, along with institutional and commercial applications ranging from hospitals using a palm scan to avoid identity mishaps to unlocking an iPhone with the touch of a finger on the screen.

These examples exhibit the common goal of seeking patterns unique to each individual, but there may be latent patterns otherwise ignored that could afford insights about groups with shared characteristics. Such discoveries might aid early diagnosis of young people who have not yet shown the symptoms of psychosis. One study has found a possible marker in genetic material from skin cells of the nose common to patients afflicted with schizophrenia.[10]

One line of research looks at finger ratios known as 2D:4D (i.e., the length of the second finger compared to the fourth, or ring, finger), which does not change in an individual's lifetime. Some studies have suggested a correspondence with sexual preference.[11] The eventual explanation could turn out to be more complex than prenatal hormonal exposure; there are cases of identical twins with different sexual preferences they express early on in life.[12]

Among future variants on innovation there is no telling what scanning technology might stumble onto—Albert Hofmann was involved in a different pharmacological pursuit when he stumbled onto LSD-25.

I think Huxley would have been fascinated by the idea of harnessing technology now used for security to gain human insights. He might have headed to his typewriter and tapped out an essay, and maybe he would have predicted such advances in his unwritten book on human resources. This was the project he and Humphry Osmond had talked about in 1963 when they met for the last time at a conference in Stockholm. Albert Hofmann, who was there too, heard Huxley describe a collaborative book intended "to address the problem of 'Human Resources,' the exploration and application of capabilities hidden in humans yet unused."[13]

Huxley's tradition of the Tuesday night circle and his exploration of nonordinary perception continues today. Some who carry on his legacy are engaged in the resurgent field of psychedelic science—researchers like Charles Grob, who has called psychedelic science and the compounds under study "of value to psychiatry, to medicine, to the human condition."[14]

One January day as I was checking a reference, and shortly before sending this manuscript to the publisher, I came upon a letter in the Huxley-Osmond correspondence that I had not noticed before. On October 22, 1963, unaware that within a month his friend would be gone, Osmond had written to Aldous, "Some exciting news about LSD-25, which seems to be one of the most efficient means of relieving pain for longer periods than we possess."

Osmond's letter raises a curious question. Aldous, in his quest for perception, entered uncharted realms and pursued unproven mind-body connections. We have assumed that his final trip was inspired by the last rites he proposed in *Island*, but perhaps this letter about pain blockage influenced why Huxley silently wrote a note requesting an intramuscular injection in his last hours.

The reason behind his choice doesn't matter, of course, and we will never know. Either way, it would have underlined his belief that moksha-medicine contains a power in this world to help alleviate suffering, it offers the potential to assist the dying, and it might even take us across a causeway between worlds.

ACKNOWLEDGMENTS

Writing about a life like Aldous Huxley's and a cultural history as colorful as psychedelics often meant deciding what, or which version, to include or set aside.

A reader may wonder about the less-familiar spelling "mescalin" instead of "mescaline," referring to the substance Humphry Osmond gave Huxley in May of 1953. I chose the shorter version as it was used in all the Huxley-Osmond correspondence.

Huxley's quest for perception gave rise to the Tuesday night salon and experiments with mescalin and LSD between the early 1950s and 1963. From this point forward I constructed a bridge to the post-2000 return of psychedelic science with a focus on the classic hallucinogens. Other authors (some of whose works are noted in the bibliography) provide details about the people and profusion of synthetics involved in the psychedelic movement between the later 1960s and today.

My childhood link to Aldous Huxley's life began with discovering long-forgotten boxes, but asking my father about their contents could only take me so far. I wish to acknowledge those who provided valuable guidance along the way, starting with John Smythies of the Center for Brain and Cognition at the University of California, San Diego, who played a key role in the 1950s flowering of psychedelic science. Dr. Smythies shared his memoirs, letters, and photographs. What I learned from our conversations provided the initial shape of this book.

Chances are the project would have stalled without the amazing support of Euphemia Osmond Blackburn, whose father is largely credited with sparking the midcentury ascent of psychedelic science. Thanks to Fee, I was able to study the original Huxley-Osmond letters as well as unpublished photographs, some of which appear in this book.

Those two personal connections came about because of William Rosar, a research associate who introduced me to his UC San Diego colleague, John Smythies. I am also most grateful to Bill for providing suggestions for early drafts of the manuscript.

I needed professional feedback when writing about the fields of psychiatry and psychedelic science, and for this I am most grateful to Edward Kaufman for his advice and critiques. I also want to thank Charles S. Grob of Harbor-UCLA Medical Center and the Los Angeles Biomedical Research Institute, who generously gave of his time both by telephone and through email. Thanks, too, to Rick Doblin of the Multidisciplinary Association for Psychedelic Studies (MAPS) for permission to use his comments from a launch event for a new area of clinical research in psychedelic medicine.

Patricia Scheifler, director of Partnership for Recovery in Sylacauga, Alabama, kindly forwarded copies of the last few letters exchanged between Osmond and Huxley. I want to thank Stephanie Starr for the captivating book-cover concept, also Lee Muhl for sharing his memories and Marc Loge for his guided tour and stories of the Los Angeles Statler Hotel in its heyday.

Speaking of guides, Anne Barry, co-owner of the Huxleys' former ranch in the Mojave Desert at Llano, showed me around the property and pointed out surviving features associated with telling moments in the Huxleys' marriage. Afterward she connected me with her neighbor Siggy Wessberg, who recalled an unusual dinner with Uncle Aldous.

A dozen years ago, around the beginning of this project, portrait artist Don Bachardy helped me arrange an interview with Laura Huxley. For that alone I am deeply grateful. Then, a decade later, he told me what it was like when Aldous Huxley sat for his portrait in 1962 and kindly gave his permission for that same drawing—which became one of the iconic images of the author—to appear in this book.

Among other permissions, I was fortunate to enjoy the cordial cooperation of Jim Spisak, Executive Director of the Aldous and Laura Huxley Literary Trust, as well as Samantha Shea at the Georges Borchardt agency. Also most helpful was Octavio Olvera of the Charles E. Young Research Library, Department of Special Collections, at UCLA.

I wish to recognize and thank the many people at Prometheus Books who gave their close and nuanced attention to the manuscript as it progressed through the detailed editing phase toward becoming a finished book, including Steven Mitchell, Peter Lukasiewicz, Catherine Roberts-Abel, Jackie Cooke, Sheila Stewart, and, for his efforts at outreach, Jake Bonar.

For the summer days I spent as a reader, I am grateful to the Huntington Library. Also to my Santa Ana College department chair, C. W. Little, who has supported this project from the start, as more recently has my SAC colleague Martin Syjuco. Additionally, I treasure the years of friendship and critiques provided by my Monday night writers group: Pam Tallman, P. J. Penman, Maralys Wills, Erv Tibbs, Terry Black, and Barbara French.

As soon as I met literary agent Dana Newman, I knew I would be in good hands. Always responsive and gracious, Dana encouraged and inspired me. In countless ways she made it happen; she found this book a home. In our literal home, my husband, Alan, stood by me and supported a sustained effort from the sidelines, overlooking bouts of crabbiness, stacks of research materials, and the pesky litter of sticky notes.

Unfortunately, I cannot thank Laura Archera Huxley, a lovely and most unusual woman, who passed away in December of 2007. Nor can I thank in any ordinary way my father, Howard, and my mother, Wanda, who raised me in a home where extraordinary perception was valued and considered worth pursuing. I wish they had lived to see *Aldous Huxley's Hands*.

NOTES

All letters from Aldous Huxley cited below, unless otherwise noted, are copyright © Laura Huxley. Reprinted by permission of Georges Borchardt, Inc., on behalf of the Aldous and Laura Huxley Trust. All rights reserved.

All letters from Humphry Osmond as cited below are printed by permission of the Humphry Osmond family.

CHAPTER 1: ADAPTATION

1. Laura Huxley, *This Timeless Moment: A Personal View of Aldous Huxley* (New York: Farrar, Straus and Giroux, 1969), p. 303.

2. Sybille Bedford, *Aldous Huxley: A Biography* (New York: Knopf, 1974), p. 47.

3. Ronald W. Clark, *The Huxleys* (New York: McGraw-Hill, 1968), p. 213.

4. Aldous Huxley, interview by Marion Brennan, Carmen Hews, Wayne Johnson, and Joe Morgan, March 18, 1957, Aldous and Laura Huxley Papers, UCLA Library Special Collections, Charles E. Young Research Library.

5. Bedford, *Aldous Huxley*, p. 2.

6. Julian Huxley, ed., *Aldous Huxley, 1894–1963: A Memorial Volume* (New York: Harper & Row, 1965), p. 98.

7. E. W. Tedlock Jr., ed., *Frieda Lawrence: The Memoirs and Correspondence* (New York: Alfred A Knopf, 1961), p. 447.

8. Miranda Seymour, *Ottoline Morrell: Life on the Grand Scale* (New York: Farrar, Straus and Giroux, 1992), p. 119.

9. Katie Roiphe, *Uncommon Arrangements: Seven Portraits of Married Life in London Literary Circles, 1910–1939* (New York: Dial Press, 2007), p. 125.

10. Seymour, *Ottoline Morrell*, p. 281.

11. Ibid., p. 1.

12. Bedford, *Aldous Huxley*, p. 80.

13. Clark, *Huxleys*, p. 214.

14. Gervas Huxley, *Both Hands: An Autobiography* (London: Chatto & Windus, 1970), p. 247.

15. Bedford, *Aldous Huxley*, p. 57.

16. Seymour, *Ottoline Morrell*, p. 157.

17. Bedford, *Aldous Huxley*, p. 91.

CHAPTER 2: SEEKERS OF PEACE

1. Aldous Huxley, "Wanted: A New Pleasure," in *Music at Night and Other Essays* (Garden City, NY: Doubleday, Doran & Co., 1931), p. 227.

2. Nicholas Murray, *Aldous Huxley: An English Intellectual* (London: Little, Brown, 2002), p. 9.

3. David King Dunaway, *Huxley in Hollywood* (New York: Harper & Row, 1989), p. 101.

4. Sybille Bedford, *Quicksands: A Memoir* (New York: Counterpoint, 2005), p. 296–98.

5. Charlotte Wolf, *Studies in Hand Reading* (London: Chatto & Windus, 1936); Aldous Huxley wrote the introduction.

6. Aldous Huxley to Mrs. Kethevan Roberts, July 30, 1936, in *Selected Letters of Aldous Huxley*, ed. James Sexton (Chicago: Ivan R. Dee, 2007), p. 7.

7. Ronald W. Clark, *The Huxleys* (New York: McGraw-Hill, 1968), p. 242.

8. Sybille Bedford, *Aldous Huxley: A Biography* (Chicago: Ivan R. Dee, 1973), p. 341.

9. Dunaway, *Huxley in Hollywood*, p. 67.

10. Bedford, *Aldous Huxley*, p. 588.

11. Christopher Isherwood, *My Guru and His Disciple* (Minneapolis, MN: University of Minnesota Press, 2001), p. 50.

12. Mary Luytens, *Krishnamurti: The Years of Fulfillment* (New York: Farrar, Straus and Giroux, 1983), p. 44.

13. Dunaway, *Huxley in Hollywood*, p. 111.

14. Ibid., p. 121.

CHAPTER 3:JACOB'S HANDS

1. Humphry Osmond, John Osmundsen, and Jerome Agel, *Understanding Understanding* (New York: Harper & Row, 1974), p. 76.

2. Franklin D. Roosevelt, "Fireside Chat 20: On the Progress of the War," February 23, 1942.

3. Huston Smith, *Cleansing the Doors of Perception* (Los Angeles: Jeremy P. Tarcher, 2000), p. 5.

4. Huston Smith, *Tales of Wonder*, with Jeffery Paine (New York: HarperCollins, 2009), p. 48.

5. Aldous Huxley, *The Art of Seeing* (New York: Harper & Brothers, 1942), p. ix.

6. Sybille Bedford, *Aldous Huxley: A Biography* (New York: Knopf, 1974), p. 719.

7. Nelda Stone, Laguna Beach Historical Society, e-mail message to author, May 25, 2009.

8. Aldous Huxley to Christopher Isherwood, July 28, 1944, in *Letters of Aldous Huxley*, ed. Grover Smith (New York: Harper & Row, 1969), pp. 510–11.

9. Bedford, *Aldous Huxley*, p. 442.

10. Christopher Isherwood, *Lost Years: A Memoir, 1945–1951*, ed. Katherine Bucknell (New York: HarperCollins Publishers, 2000), p. 273.

CHAPTER 4: CHASING SCHIZOPHRENIA

1. Jonathan Engel, *American Therapy* (New York: Gotham Books, 2009), p. 152.

2. Ibid., p. xii.

3. John Smythies and Vanna Smythies, *Two Coins in a Fountain.* (Charleston, NC: Book Surge, 2005), p. 29.

4. Smythies and Smythies, *Two Coins in A Fountain*, p. 25.

5. Aldous Huxley, *The Doors of Perception* (New York: Harper & Brothers, 1954), p. 9.

6. Heffter Research Institute, "About: Dr. Arthur Heffter," http://www.heffter.org/about-arthurheffter.htm (accessed August 18, 2015).

7. Humphry Osmond, John A. Osmundsen, and Jerome Agel, *Understanding Understanding* (New York: Harper & Row, 1974), p. 77.

8. Ibid.

9. John Smythies, "The Extension of Mind: A New Theoretical Basis for psi Phenomenon," *Journal of the Society for Psychical Research*, 36 (September–October 1951).

10. H. H. Price to John Smythies, July 17, 1952, John R. Smythies Papers, UAB Archives, University of Alabama at Birmingham.

11. C. G. Jung to John Smythies, February 4, 1952, John R. Smythies Papers, UAB Archives, University of Alabama at Birmingham.

12. Aldous Huxley to John Smythies, Nov. 25, 1952, John R. Smythies Papers, UAB Archives, University of Alabama at Birmingham.

13. John Smythies, in discussion with the author, February 26, 2007.

14. Smythies and Smythies, *Two Coins in the Fountain*, p. 37.

15. Harold Cummins to Howard Thrasher, October 2, 1952.

16. Jonathan Engel, *American Therapy* (New York: Gotham Books, 2009), p. 44–45.

17. Humphry Osmond and John Smythies, "The Present State of Psychological Medicine," *Hibbert Journal: A Quarterly Review of Religion, Theology, and Philosophy*, 51, no. 201 (October 1952–July 1953): 134–42.

18. Osmond, Osmundsen, and Agel, *Understanding Understanding*, p. 55.

19. Ibid., p. 55.

20. Fee Osmond Blackburn, in telephone discussion with the author, February 26, 2007.

CHAPTER 6: TUESDAY NIGHTS ON NORTH KINGS ROAD

1. Aldous Huxley to J. B. Rhine, May 11, 1943, in *Letters of Aldous Huxley*, ed. Grover Smith (New York: Harper & Row, 1969), p. 489.

2. David King Dunaway, *Aldous Huxley Recollected: An Oral History* (Walnut Creek, CA: AltaMira Press, 1999), p. 115.

3. Sybille Bedford, *Aldous Huxley: A Biography* (Chicago: Ivan R. Dee, 1973), p. 498–99.

4. Leslie LeCron, "The Paranormal in Hypnosis," *Tomorrow* (Spring 1955).

5. Roy Maypole was mentioned as a soldier and former radio soap-opera impresario in *Time*, May 17, 1944, http://www.time.com/time/magazine/article/0,9171,850874,00.html (accessed January 4, 2015).

6. Jean Dunn, ed. *Consciousness and the Absolute: The Final Talks of Sri Nisargadatta Maharaj* (Durham, NC: Acorn Press, 1994)., http://www.weare sentience.com/jean-dunn–nisargadatta-maharaj.html (accessed January 4, 2015).

7. Aldous Huxley, "Distractions," in *Vedanta for the Western World*, ed. Christopher Isherwood (New York: Viking Press, 1945), p. 127.

8. Bedford, *Aldous Huxley*, p. 509.

9. Aldous Huxley to Humphry Osmond, December 17, 1953.

10. Aldous Huxley to Julian Huxley, January 25, 1953, in Smith, *Letters of Aldous Huxley*.

CHAPTER 7: TAKE A SIP OF AMAZEMENT

1. Humphry Osmond to Aldous Huxley, April 19, 1953.

2. The description of Osmond's first mescalin experience was later published in Humphry Osmond, John A. Osmundsen, and Jerome Agel, *Understanding Understanding* (New York: Harper & Row, 1974), p. 83.

3. Aldous Huxley to Humphry Osmond, April 10, 1953.

4. H. H. Price to John Smythies, April 12, 1953, John R. Smythies Papers, UAB Archives, the University of Alabama at Birmingham.

5. Humphry Osmond to Aldous Huxley, April 13, 1953.

6. Humphry Osmond to Aldous Huxley, April 24, 1953.

7. Charles S. Grob, ed., *Hallucinogens: A Reader* (New York: Jeremy P. Tarcher, 2002), p. 267.

8. Humphry Osmond to Aldous Huxley, March 31, 1953.

9. Stacy Horn, *Unbelievable: Investigations into Ghosts, Poltergeists, Telepathy, and other Unseen Phenomena, from the Duke Parapsychology Laboratory* (New York: HarperCollins, 2009), p. 48.

10. Ibid., p. 46.

11. Humphry Osmond to Jane Osmond, May 6, 1953.

12. Aldous Huxley, *The Doors of Perception* (New York: Harper & Row, 1954), p. 14.

13. Ibid., p. 16.

14. Ibid., p. 17.

15. Ibid., p. 21.

16. Ibid., p. 22.

17. Ibid., p. 28.

18. Ibid., p. 57.

19. Ibid., p. 50.

20. Ibid., p. 17.

CHAPTER 8: MESCALIN AND MARILYN MONROE

1. Albert Hofmann, *LSD: My Problem Child* (Los Angeles: J. P. Tarcher, 1983), p. 15.

2. Ibid., p. 36.

3. Ibid., p. 51.

4. Humphry Osmond to Aldous Huxley, May 25, 1953.

5. John Smythies and Vanna Smythies, *Two Coins in a Fountain* (Charleston, SC: Book Surge, 2005), p. 54.

6. Aldous Huxley to Julian Huxley, January 25, 1953.

7. Hofmann, *LSD: My Problem Child*, p. 58.

8. John Smythies, in discussion with the author, February 26, 2007.

9. Erika Dyck, *Psychedelic Psychiatry: LSD From Clinic to Campus* (Baltimore: Johns Hopkins University Press, 2008), p. 52.

10. It is uncertain which individuals on the list of prospects Osmond and Smythies, or possibly even Huxley, had specifically invited in this early planning stage.

11. Humphry Osmond to Aldous Huxley, June 21, 1953.

12. Humphry Osmond to Aldous Huxley, June 26, 1953.

13. Letter from H. H. Price to John Smythies, July 20, 1953, John R. Smythies Papers, UAB Archives, the University of Alabama at Birmingham.

14. Aldous Huxley to Humphry Osmond, August 17, 1953.

15. Humphry Osmond to Aldous Huxley, August 22, 1953.

16. Aldous Huxley to Humphry Osmond, September 26, 1953.

17. Stacy Horn, *Unbelievable: Investigations into Ghosts, Poltergeists, Telepathy, and Other Unseen Phenomenon, from the Duke Parapsychology Lab* (New York: HarperCollins, 2009), p. 128.

18. Humphry Osmond, review of *The Doors of Perception*, by Aldous Huxley, unpublished typescript draft. (A version later appeared in *Tomorrow*, Spring 1954.)

CHAPTER 9: SLAMMING THE DOORS

1. Aldous Huxley to Humphry Osmond, January 25, 1954.

2. Humphry Osmond to Aldous Huxley, April 16, 1954.

3. Aldous Huxley to Harold Raymond, March 8, 1954.

4. Berton Roueché, "Shimmering Hours," *New York Times*, February 7, 1954, BR6.

5. C. D. Broad to John Smythies, October 12, 1955, John R. Smythies Papers, UAB Archives, the University of Alabama at Birmingham.

6. John Smythies, in discussion with the author, February 26, 2007.

7. John Smythies to Aldous Huxley, February 5, 1955, John R. Smythies Papers, UAB Archives, the University of Alabama at Birmingham.

8. Sybille Bedford, *Aldous Huxley: A Biography* (Chicago: Ivan R. Dee, 1973), p. 427.

9. An example is S. Schmidt, R. Schneider, J. Utts, and H. Walach, "Distant Intentionality and the Feeling of Being Stared At: Two Meta-Analyses," *British Journal of Psychology*, 95, no. 2 (2004): 235–47.

CHAPTER 10: THE FRENCH CONNECTION

1. Aldous Huxley to Humphry Osmond, April 9, 1954.

2. Aldous Huxley to Clifford Bax, April 10, 1954.

3. Humphry Osmond to Aldous Huxley, April 23, 1954.

4. Ibid.

5. H. H. Price, who met Huxley at Eileen's symposium that April, had recommended Ducasse for the Outsight group, as well as A. J. Ayer, Gilbert Ryle, W. T. Stace, and Gabriel Marcel. (H. H. Price to John Smythies, July 20, 1953.)

6. Humphry Osmond to Aldous Huxley, June 12, 1954.

CHAPTER 11: ALDOUS GOES ROGUE

1. Humphry Osmond to Aldous Huxley, July 11, 1954.

2. Ibid.

3. Humphry Osmond to Aldous Huxley, August 22, 1954.

4. Humphry Osmond to Aldous Huxley, October 29, 1954.

5. Humphry Osmond to Aldous Huxley, April 23, 1954.

6. Aldous Huxley to Humphry Osmond, October 16, 1954.

7. Humphry Osmond to Aldous Huxley, July 11, 1954.

8. Sybille Bedford, *Aldous Huxley: A Biography* (Chicago: Ivan R. Dee, 1973), p. 547.

9. Humphry Osmond to Aldous Huxley, December 7, 1954.

10. Humphry Osmond to Aldous Huxley, December 24, 1954.

11. Humphry Osmond to Aldous Huxley, December 7, 1954.

12. Erika Dyck, *Psychedelic Psychiatry: LSD from Clinic to Campus* (Baltimore: Johns Hopkins University Press, 2008), p. 90.

13. Aldous Huxley to Humphry Osmond, January 12, 1955.

CHAPTER 12: MARIA

1. Aldous Huxley to Humphry Osmond, February 10, 1955.

2. Humphry Osmond to Aldous Huxley, February 27, 1955.

3. Aldous Huxley to Humphry Osmond, March 18, 1955.

CHAPTER 13: CONSTERNATION AT MENNINGER AND MIRAMAR

1. Humphry Osmond to Aldous Huxley, April 14, 1955.

2. Humphry Osmond to Aldous Huxley, April 23, 1954.

3. Humphry Osmond to Aldous Huxley, April 14, 1955.

4. Humphry Osmond to Aldous Huxley, May 13, 1955.

5. Humphry Osmond to Aldous Huxley, June 2, 1955.

6. Aldous Huxley to Humphry Osmond, June 18, 1955.

7. The study would not be published for twenty years. (J. A. Gengerelli and H. Thrasher, "Termination of Palmar Main Line of the Left Hand in Relation to Mental Pathology," *The Journal of Psychology: Interdisciplinary and Applied* 101, no. 2 (1979): 279–86.)

CHAPTER 14: SUCCESSOR WIFE

1. Laura Huxley, *This Timeless Moment: A Personal View of Aldous Huxley* (New York: Farrar, Straus and Giroux, 1968), p. 3.

2. Ibid., p. 17.

3. Sybille Bedford, *Aldous Huxley: A Biography* (Chicago: Ivan R. Dee, 1973), p. 595.

4. Ibid., p. 596.

5. Huxley, *This Timeless Moment*, p. 27.

CHAPTER 15: TAKE A PINCH OF PSYCHEDELIC

1. Aldous Huxley to Humphry Osmond, March 30, 1956.

2. Ibid.

3. Humphry Osmond to Aldous Huxley, ca. April 2, 1956.

CHAPTER 16: CONVERGENCE IN CAMBRIDGE

1. Humphry Osmond, John A. Osmundsen, and Jerome Agel, *Understanding Understanding* (New York: Harper & Row, 1974), p. 95.

2. Huston Smith, *Cleansing the Doors of Perception* (Los Angeles: Jeremy P. Tarcher, 2000), p. 151.

3. Osmond, Osmundsen, and Agel, *Understanding Understanding*, p. 95.

4. Humphry Osmond to Aldous Huxley, April 23, 1954.

5. Robert Greenfield, *Timothy Leary: A Biography* (Orlando: Harcourt, 2006), pp. 148–53.

6. Ibid., pp. 180–84.

7. For example, Rick Doblin, "Dr. Leary's Concord Prison Experiment: A 34 Year Follow-Up Study," *Journal of Psychoactive Drugs*, 30, no. 4 (1998): 10–18.

8. Albert Hofmann, *LSD: My Problem Child* (Los Angeles: J. P. Tarcher, 1983), p. 134.

CHAPTER 17: PURIFYING FIRE

1. Humphry Osmond to Aldous Huxley, February 20, 1961.

2. Aldous Huxley to Eileen Garrett, May 19, 1961.

3. John Smythies, in discussion with the author, February 27, 2007.

4. Aldous Huxley, interview by Marion Brennan, Carmen Hews, Wayne Johnson, and Joe Morgan, March 18, 1957, Aldous and Laura Huxley Papers, UCLA Library Special Collections, Charles E. Young Research Library.

5. Sybille Bedford, *Aldous Huxley: A Biography* (Chicago: Ivan R. Dee, 1973), p. 731.

6. Ibid., p. 737.

7. Julian Huxley, ed. *Aldous Huxley, 1894–1963: A Memorial Volume* (New York: Harper & Row, 1965), pp. 157–58. Essay by Christopher Isherwood, copyright © 1965 by Christopher Isherwood, collected in *Aldous Huxley, 1894–1963: A Memorial Volume*, used by permission of The Wylie Agency LLC.

CHAPTER 18: BACKPEDAL ON LSD

1. Laura Huxley, *This Timeless Moment: A Personal View of Aldous Huxley* (New York: Farrar, Straus and Giroux, 1968), p. 302.

2. Sylvia Jukes Morris, "Clare, In Love and War," *Vanity Fair*, July 2014.

3. Susan Carpenter, "Oscar Janiger: Pioneer in Psychedelic Research," *Los Angeles Times*, August 17, 2001, p. B15.

4. Rick Doblin, Jerome E. Beck, Kate Chapman, and Maureen Alioto, "Dr. Oscar Janiger's Pioneering LSD Research: A Forty Year Follow-up," *Bulletin of the Multidisciplinary Association for Psychedelic Studies*, 9, no. 1 (Spring 1999).

5. Aldous Huxley to Humphry Osmond, October 16, 1954.

6. Bob Byers, "Stanford Reveals CIA Links," *New Scientist*, October 13, 1977, p. 81.

7. "Dr. Sidney Cohen, 76, Dead: Studied Mood-Altering Drugs," *New York Times*, May 17, 1987, http://nytimes.com/1987/05/17/obituaries/dr-sidney-cohen-76-dead-studied-mood-altering-drugs.html (accessed December 2, 2014).

8. Humphry Osmond to Aldous Huxley, April 24, 1953.

9. Robert Greenfield, *Timothy Leary: A Biography* (Orlando: Harcourt, 2006), p. 195.

10. "1964: The Year We Stood Up and Split Apart," *The American Experience*, http://www.pbs.org/wgbh/americanexperience/films/1964 (accessed November 1, 2014).

11. "SFSU Centennial History: Long Narrative of SF State," *San Francisco State University, 1899–1999*, March 20, 2009, http://ww.sfsu.edu/~100years/history/long.htm (accessed August 18, 2015).

12. Richard Nixon, "Special Message to the Congress on Drug Abuse Prevention and Control," June 17, 1971, *American Presidency Project*, http://www.presidency.ucsb.edu/ws/index.php?pid=3048 (accessed November 2014).

CHAPTER 20: THE VISIONARY EXPERIENCE

1. Don Bachardy, in discussion with the author, January 13, 2015.

2. Sybille Bedford, *Aldous Huxley: A Biography* (Chicago: Ivan R. Dee, 1973), p. 715.

3. Charles Grob, in discussion with the author, August 8, 2014.

4. Aldous Huxley, "Visionary Experience" lecture at Monterey Peninsula College, April 29, 1962, broadcast on KPFA, December 1, 1968, Pacifica Radio Archives BB1671.

CHAPTER 21: THE RETURN OF PSYCHEDELIC SCIENCE

1. "Psychedelic Science: From '60s Counterculture to Modern Medicine," public event at Crawford Family Forum, Pasadena, CA, September 15, 2014, http://scpr.org/events/2014/09/15/1506/psychedelic-science/.

2. Examples of psychological measures include Beck Depression Inventory (BDI), Profile of Mood States (POMS), State-Trait Anxiety Inventory (STAI), and 5D-ASC, or 5-Dimensional Altered State of Consciousness Profile.

3. Pilot study in Zurich of LSD-assisted psychotherapy with twelve patients with advanced cancer was conducted from 2008 to 2011 under lead physician Peter Gasser. (Benedict Carey, "LSD, Reconsidered for Therapy," *New York Times*, March 3, 2014, http://www.nytimes.com/2014/03/04/health/lsd-reconsidered-for-therapy.html?_r=0.)

4. A pilot study through the Beckley Foundation–Imperial College

Research Program looked at how LSD works on the human brain to produce its characteristic psychological effects (www.beckleyfoundation.org/2014/09/frequently-asked-questions-beckley-imperial-college-research-programme/ [accessed July 31, 2015]).

5. Charles Grob, "Appendix," in *Hallucinogens: A Reader*, ed. Charles Grob (New York: Jeremy P. Tarcher, 2002), p. 219–20.

6. C. S. Grob et al., "Psychobiologic Effects of 3,4-methylenedioxymethampetamine (MDMA) in Humans: Methodological Considerations and Preliminary Data," *Behavioral Brain Research*, n. 73 (1996), pp. 103–107.

7. Charles S. Grob, "The Use of Psilocybin in Patients with Advanced Cancer and Existential Anxiety," in M. Winkelman and T. Roberts, eds., *Psychedelic Medicine: New Evidence for Hallucinogens as Treatments*, vol. 1 (Westport, CT: Praeger/Greenwood, 2007), pp. 205–216.

8. Charles Grob, in discussion with the author, August 8, 2014.

9. Stanislav Grof and Joan Halifax, *The Human Encounter with Death* (New York: E. P. Dutton, 1977), p. 16.

10. "Great Entactogen-Empathogen Debate," Letters to the Editor, *MAPS Newsletter*, 4, no. 2 (Summer 1993), www.maps.org/news-letters/v04n2/0424eed .html (accessed November 2, 2014).

11. Sanjay Gupta, "Ecstasy as a Possible Treatment for Severe PTSD," CNN, December 1, 2012, http://www.cnn.com/2012/12/01/health/ecstasy-ptsd-1/.

12. Linda Marsa, "Could an Acid Trip Cure Your OCD?" *Discover*, May 16, 2008, http://www.discovermagazine.com/2008/jun/16-could-an-acid-trip-cure -your-ocd (accessed November 7, 2014).

13. Rick Strassman, "Sitting for Sessions: Dharma & DMT Research," in *Hallucinogens*, p. 113.

14. Charles Grob, "The Psychology of Ayahuasca," in *Hallucinogens*, p. 195.

15. R. R. Griffiths, et al., "Mystical-Type Experiences Occasioned by Psilocybin Mediate the Attribution of Personal Meaning and Spiritual Significance 14 Months Later," *Journal of Psychopharmacology* 22, no. 6 (August 2008): 621–32.

16. Eric Kast and Vincent Collins, "LSD Used as Analgesic," *Journal of the American Medical Association* 187, no. 1 (1964): 33. (The abstract said of the drug, "Heretofore used only in psychic experiments and in studies of schizophrenia. . . .")

17. C. S. Grob, et al, "Pilot Study of Psilocybin Treatment for Anxiety in Patients with Advanced Stage Cancer," *Archives of General Psychiatry* 68, no. 1 (2011): 71–78.

18. F. A. Moreno, et al., "Safety, Tolerability, and Efficacy of Psilocybin in 9 Patients with Obsessive-Compulsive Disorder," *Journal of Clinical Psychiatry* 67, no. 11 (2006): 1735–40, www.ncbi.nlm.nih.gov/pubmed/17196053?dopt =AbstractPlus.

19. On its website, the Heffter Institute cites results of the 2006 pilot study with psilocybin as showing "a potential breakthrough treatment."

20. Alicia L. Danforth, et al., "MDMA-Assisted Therapy: A New Treatment Model for Social Anxiety in Autistic Adults," *Progress in Neuropsychopharmacology and Biological Psychiatry*, March 25, 2015, www.sciencedirect.com/science/ article/pii/S0278584615000603 (accessed August 1, 2015).

21. Bob Morris. "Ayahuasca: A Strong Cup of Tea," *New York Times*, June 13, 2004, p. 2.

22. Charles Grob, in discussion with the author, August 8, 2014.

23. Rick Doblin, comment at event launching MAPS-sponsored adult autism study, Culver City, CA, November 13, 2014.

24. The Beckley Foundation investigates the brain mechanisms underlying psychoactive substances, such as cannabis and psychedelics (http://www.beckley foundation.org/2014/09/frequently-asked-questions-beckley-imperial-college -research-programme/ [accessed January 5, 2015]).

25. Robin Carhart-Harris, Mendel Kaele and David Nutt, "How Do Hallu- cinogens Work on the Brain?" *The Psychologist* 27, no. 9 (September 2014): 662.

26. Mo Constandi, "Psychedelic Chemical Subdues Brain Activity," *Nature*, January 23, 2012, http://www.nature.com/news/psychedelic-chemical -subdues-brain-activity-1.9878 (accessed November 5, 2014).

CHAPTER 22: BESIDE PSYCHEDELICS

1. Sharon Begley, "Scans of Monks' Brains Show Meditation Alters Structure, Functioning," *Wall Street Journal*, November 5, 2004, http://www.wsj .com/articles/SB109959818932165108.

2. Andrew B. Newberg, interview, *Talk of the Nation*, National Public Radio, July 14, 2011.

3. Matt Danzico, "Brains of Buddhist Monks Scanned in Meditation Study," BBC News, April 24, 2011, http://www.bbc.co.uk/news/world-us -canada-12661646 (accessed August 10, 2014).

4. Sarah Knapton, "First Hint of 'Life after Death' in Biggest Ever Scientific Study," *Telegraph*, October 7, 2014, http://www.telegraph.co.uk/news/science/science-news/11144442/First-hint-of-life-after-death-in-biggest-ever-scientific-study (accessed November 10, 2014).

5. Rick Doblin, comment at event launching MAPS-sponsored adult autism study, November 13, 2014, Culver City, CA. (Article in press: Alicia Danforth, et al., "MDMA-Assisted Therapy: A New Treatment Model for Social Anxiety in Autistic Adults," *Progress in Neuro-Psychopharmacology and Biological Psychiatry*.)

6. Stanislaw Grof and Christina Grof, *Holotropic Breathwork: A New Approach to Self-Exploration and Therapy* (Albany: State University of New York, 2010).

7. Aldous Huxley, *Heaven and Hell* (New York: Harper & Brothers, 1956), p. 12.

8. Aldous Huxley, *Those Barren Leaves* (New York: Harper and Bros., 1925), p. 342.

9. Natasha Singer, "When a Palm Reader Knows More Than Your Life Line," *New York Times*, November 11, 2012, p. 3.

10. Daniel Akst. "A Nose for Mental Illness," *Wall Street Journal*, May 11, 2013.

11. Deborah Blum, "Is Our Fate Written in the Lengths of Our Fingers?" *Los Angeles Times*, July 8, 2001, p. M3.

12. "The Science of Sexual Orientation," *60 Minutes*, CBS, March 12, 2006.

13. Albert Hofmann, *LSD: My Problem Child* (Los Angeles: J. P. Tarcher, 1983), p.175.

14. Charles Grob, comment at "Psychedelic Science: From '60s Counter-culture to Modern Medicine," public event at Crawford Family Forum, Pasadena, CA, September 15, 2014, http://scpr.org/events/2014/09/15/1506/psychedelic-science/.

BIBLIOGRAPHY

Bedford, Sybille. *Aldous Huxley: A Biography*. Chicago: Ivan R. Dee. 1973.

———. *Quicksands: A Memoir*. New York: Counterpoint, 2005.

Brown, David Jay. *The New Science of Psychedelics: At the Nexus of Culture, Consciousness, and Spirituality*. Rochester, VT: Park Street Press, 2013.

Clark, Ronald W. *The Huxleys*. New York: McGraw Hill, 1968.

Doblin, Rick, and Brad Burgee. *Manifesting Minds: A Review of Psychedelics in Science, Medicine, Sex and Spirituality*. Berkeley, CA: Evolver Editions, 2014.

Dunaway, David King. *Huxley in Hollywood*. New York: Harper & Row, 1989.

———. *Aldous Huxley Recollected: An Oral History*. Walnut Creek, CA: Alta Mira Press, 1999.

Dyck, Erika. *Psychedelic Psychiatry: LSD From Clinic to Campus*. Baltimore: Johns Hopkins University Press, 2008.

Engel, Jonathan. *American Therapy*. New York: Gotham Books, 2009.

Greenfield, Robert. *Timothy Leary: A Biography*. Orlando, FL: Harcourt, 2006.

Grob, Charles S., ed. *Hallucinogens: A Reader*. New York: Jeremy P. Tarcher, 2002.

Hagerty, Barbara Bradley. *Fingerprints of God: The Search for the Science of Spirituality*. New York: Riverhead Books, 2009.

Heard, Gerald. *Pain, Sex and Time: A New Outlook on Evolution and the Future of Man*. New York: Harper and Brothers, 1939.

Hofmann, Albert. *LSD: My Problem Child*. Los Angeles: J. P. Tarcher, 1983.

Horn, Stacy. *Unbelievable: Investigations into Ghosts, Poltergeists, Telepathy, and other Unseen Phenomena, from the Duke Parapsychology Laboratory*. New York: HarperCollins, 2009.

Horowitz, Michael, and Cynthia Palmer, eds. *Moksha: Aldous Huxley's Classic Writings on Psychedelics and the Visionary Experience*. Rochester, VT: Park Street Press, 1999.

Huxley, Aldous. *After Many a Summer Dies the Swan*. New York: Harper Colophon, 1983.

———. *Antic Hay*. New York: Random House Modern Library, 1932.

———. *Ape and Essence*. New York: Bantam, 1958.

———. *Brave New World*. New York: Harper Perennial, 2006.

————. *Brief Candles*. London: Chatto and Windus, 1930.

————. *Crome Yellow*. London: Chatto and Windus, 1921.

————. *Eyeless in Gaza*. London: Chatto and Windus, 1936.

————. *Grey Eminence*. New York: Harper Colophon, 1966.

————. *Heaven and Hell*. New York: Harper & Brothers, 1956.

————. *Island*. New York: Harper & Row, 1962.

————. *Mortal Coils*. New York: George H. Doran, 1922.

————. *Music at Night, and Other Essays*. Garden City, NY: Doubleday, Doran, 1931.

————. *Point Counter Point*. New York: Doubleday, Doran, 1928.

————. *Texts and Pretexts*. Westport, CT: Greenwood Press, 1976.

————. *The Art of Seeing*. New York: Harper & Brothers, 1942.

————. *The Devils of Loudun*. New York: Harper & Brothers, 1952.

————. *The Doors of Perception*. New York: Harper & Brothers, 1954.

————. *The Genius and the Goddess*. London: Granada Publishing, 1982.

————. *The Perennial Philosophy*. New York: Harper & Brothers, 1944.

————. *Those Barren Leaves*. New York: George H. Doran, 1925.

————. *Time Must Have a Stop*. Champagne, IL: Dalkey Archive Press, 1998.

————. *Tomorrow and Tomorrow and Tomorrow, and Other Essays*. New York: Harper & Brothers, 1952.

Huxley, Aldous, and Christopher Isherwood. *Jacob's Hands: A Fable*. New York: St. Martin's Griffin, 1998.

Huxley, Gervas. *Both Hands: An Autobiography*. London: Chatto & Windus, 1970.

Huxley, Julian, ed. *Aldous Huxley, 1894–1963: A Memorial Volume*. Harper & Row, 1965.

Huxley, Laura. *This Timeless Moment: A Personal View of Aldous Huxley*. New York: Farrar, Straus and Giroux, 1968.

Isherwood, Christopher, ed. *Diaries, Volume One: 1939–1960*. Katherine Bucknell, ed. New York: HarperCollins, 1996.

————. *Lost Years: A Memoir, 1945–1951*. Katherine Bucknell, ed. New York: HarperCollins, 2000.

James, William. *The Varieties of Religious Experience*. New York: New American Library, 1958.

Lattin, Don. *The Harvard Psychedelic Club*. New York: HarperCollins, 2010.

————. *Distilled Spirits: Getting High, Then Sober, With a Famous Writer, a Forgotten Philosopher, and a Hopeless Drunk*. Berkeley, CA: University of California Press, 2012.

Lee, Martin, and Bruce Shlain. *Acid Dreams: The Complete Social History of LSD: The CIA, the Sixties, and Beyond*. New York: Grove/Atlantic, 1992.

Luytens, Mary. *Krishnamurti: The Years of Fulfillment*. New York: Farrar, Straus and Giroux, 1983.

Maxtone-Graham, John. *Normandie*. New York: W. W. Norton, 2009.

Morrell, Ottoline. *Ottoline at Garsington: Memoirs of Lady Ottoline Morrell, 1915–1918*. Robert Gathorne Hardy, ed. New York: Alfred A. Knopf, 1975.

Murray, Nicholas. *Aldous Huxley: An English Intellectual*. London: Little Brown, 2002.

Osmond, Humphry, John A. Osmundsen, and Jerome Agel. *Understanding Understanding*. New York: Harper & Row, 1974.

Roach, Mary. *Spook: Science Tackles the Afterlife*. New York: W. W. Norton, 2005.

Roiphe, Katie. *Uncommon Arrangements: Seven Portraits of Married Life in London Literary Circles, 1910–1939*. New York: Dial Press, 2007.

Sawyer, Dana. *Aldous Huxley: A Biography*. New York: Crossroad, 2002.

Seymour, Miranda. *Ottoline Morrell: Life on the Grand Scale*. New York: Farrar, Straus and Giroux, 1992.

Sexton, James. *Selected Letters of Aldous Huxley*. Chicago: Ian R. Dee, 2007.

Shroder, Tom. *Acid Test: LSD, Ecstasy, and the Power to Heal*. New York: Blue Rider Press, 2014.

Smith, Grover, ed. *Letters of Aldous Huxley*. New York: Harper & Row, 1969.

Smith, Huston. *Cleansing the Doors of Perception*. Los Angeles: Jeremy P. Tarcher, 2000.

———. *Tales of Wonder*. With Jeffery Paine. New York: HarperCollins, 2009.

———. *The Religions of Man*. Harper Colophon, 1964.

Smythies, John, and Vanna Smythies. *Two Coins in a Fountain*. Charleston, SC: Book Surge, 2005.

Tedlock, E. W., Jr. *Frieda Lawrence: The Memoirs and Correspondence*. New York: Alfred A. Knopf, 1964.

Thrasher, Howard. *Aircraft Lofting and Template Design*. San Francisco: Aviation Press, 1942.

Thurman, Robert A. F., trans. *The Tibetan Book of the Dead*. New York: Bantam, 1994.

INDEX